AMERICAN MATHEMATICAL SOCIETY
TRANSLATIONS

Series 2

Volume 134

Fourteen Papers Translated from the Russian

by

A. Ya. Aĭzenshtat
V. V. Baranov
B. M. Bil′man
B. K. Boguta
V. A. Bondarenko
G. F. Kushner
V. I. Malykhin
Yu. V. Mel′nichuk
B. V. Novikov
N. S. Podtsykin
B. È. Shapirovskiĭ
V. V. Sharko
V. P. Soltan
G. D. Suvorov
A. V. Zarelua

American Mathematical Society
Providence, Rhode Island

Translation edited by BEN SILVER

1980 *Mathematics Subject Classification* (1985 *Revision*). Primary 11K60, 20-02, 20MXX, 26A51, 30-02, 30D40, 30G15, 45E05, 52A07, 53CXX, 54-02, 54A25, 54CXX, 54DXX, 54E35, 57RXX, 60J10, 65V05, 93E20; Secondary 03E50, 04A30, 08B15, 17BXX, 30C60, 30E20, 46B05, 54E65.

Library of Congress Cataloging-in-Publication Data

Fourteen papers translated from the Russian.

(American Mathematical Society translations, ISSN 0065-9290; ser. 2, v. 134)

1. Mathematics–Collected works–Translations from Russian. I. Aĭzenshtat, A. I͡A. (Anna I͡Akovlevna). II. Series.

QA3.A572 ser. 2, vol. 134 510 s [510] 86-32162

ISBN 0-8218-3110-0

Contents

Russian Contents*

*The American Mathematical Society scheme for transliteration of Cyrillic may be found at the end of index issues of *Mathematical Reviews*.

Amer. Math. Soc. Transl.
(2) Vol. **134**, 1987

Hausdorff Dimension in Diophantine Analysis on Curves

UDC 511.36

YU. V. MEL'NICHUK

Let the functions $f_1(x), \dots, f_n(x)$ be defined and n times continuously differentiable on $[a, b]$, and let the Wronskian $W(f_1'(x), \dots, f_n'(x)) \neq 0$ at the points $x \in [a, b]$. In $\mathbf{R}^n$, $n \geq 2$, consider the curve $\Gamma = (f_1(x), \dots, f_n(x))$ and denote by $A_n(w)$ the set of points $x \in [a, b]$ at which the inequality

$$|a_0 + a_1 f_1(x) + \cdots + a_n f_n(x)| < H^{-w}, \tag{1}$$

where $H = \max(|a_0|, \dots, |a_n|)$, has infinitely many solutions in rational integers $a_0, a_1, \dots, a_n$. If for every $\varepsilon > 0$ the Lebesgue measure of the set $\mathcal{M}(\nu_n + \varepsilon)$ is zero, the curve Γ is called a *ν-extremal* [1]. Under the conditions on $f_i(x)$ described above, the $(n + n^{-1} - 1)$-extremality of Γ was established in [2].

For $w_1 < w_2$ we have $A_n(w_1) \supset A_n(w_2)$, but for a ν-extremal curve the sets $A_n(w_1)$ and $A_n(w_2)$ with $w_1 > \nu_n$ have the same Lebesgue measure, namely zero. Therefore in order to study them further we turn to Hausdorff measure.

The Hausdorff measure of a set $\mathcal{M}$ is a real number, denoted by $\dim \mathcal{M}$, the infimum of numbers ρ such that for every $\delta > 0$ there is a countable set $\mathcal{H}(\delta, \rho)$ of intervals $I_1, I_2, \dots$ that cover $\mathcal{M}$ and satisfy the following conditions:

a) $\operatorname{meas} I_j < \delta$, $j = 1, 2, \dots$, where $\operatorname{meas} I_j$ is the length of I_j.

b) $\sum_{j=1}^{\infty} (\operatorname{meas} I_j)^{\delta} < \infty$.

THEOREM 1. *The Hausdorff dimension of $A_n(w)$ satisfies the inequality*

$$\dim A_n(w) \leq (n^2 + n)(w + 1)^{-1}.$$

PROOF. Denote $a_0 + a_1 f_1(x) + \cdots + a_n f_n(x)$ by $F(x)$. For $F(x)$ with given $a_0, a_1, \dots, a_n$ we consider the set $\sigma_F = \{x \in [a, b]\colon |F(x)| < H^{-w}\}$.

1980 *Mathematics Subject Classification* (1985 *Revision*). Primary 11K60.

Translation of Studies in the Qualitative Theory of Differential Equations and Its Applications (V. Ya. Skorobogat'ko, editor), Inst. Mat. Akad. Nauk Ukrain. SSR, Kiev, 1978, pp. 35–38; MR **82j**:10087.

It follows from [2] that σ_F consists of a finite number, depending only on n and the length of the interval $[a, b]$, of intervals whose total length does not exceed $H^{-w+1/n}$. There are not more than $3H^n$ different functions $F(x)$ with the same H. The series that appears in the definition of $\dim A_n(w)$ is, up to a constant, majorized by the series

$$\sum_{H=1}^{\infty} (H^{-(w+1)/n})^{\rho} \cdot H^n.$$

It converges for $\rho > (n^2 + n)(w + 1)^{-1}$, as Theorem 1 shows.

THEOREM 2. *If Γ is ν-extremal, the Hausdorff dimension of $A_n(w)$ satisfies the inequality*

$$\dim A_n(w) \geq (1 + (\nu + n^{-1} - 1)^{-1})(w + 1)^{-1}.$$

COROLLARY 1. *For a 1-extremal curve*

$$\dim A_n(w) \geq (n + 1)(w + 1)^{-1}.$$

COROLLARY 2. *For a ν-extremal curve*

$$\dim A_n(w) \geq (w + 1)^{-1} \max(3, 1 + (\nu + n^{-1} - 1)^{-1}).$$

Let Γ be ν-extremal; fix a real number ε, $0 < \varepsilon < 1$, and set $\delta = \varepsilon/2$. Consider an arbitrary interval $\mathcal{I} \subset [a, b]$. Let $R(\mathcal{I}, H)$ denote the set of points $\xi \in \mathcal{I}$ for which there is a zero α of some function $F(x) = a_0 + a_1 f_1(x) + \cdots + a_n f_n(x)$ with $\max(|a_0|, \ldots, |a_n|) = H$ such that

$$|\xi - \alpha| < H^{-1-(1-\varepsilon)(\nu+n^{-1}-1)^{-1}}.$$

LEMMA 1. *As H increases the measure of $R(\mathcal{I}, H)$ tends to the measure of $\mathcal{I}$.*

PROOF. Let $\xi \in \mathcal{I}$ and $\xi \notin R(\mathcal{I}, H)$ for every H. Choose a sufficiently large number H_0. By Minkowski's theorem ([3], p. 151) the system of inequalities

$$\begin{aligned} |a_0 + a_1 f_1(\xi) + \cdots + a_n f_n(\xi)| &< H_0^{-\nu_n + (n-1)\delta}; \\ |a_1 f_1'(\xi) + \cdots + a_n f_n'(\xi)| &< c H_0^{\nu_n + 1 - n}; \\ |a_2| &< H_0^{1-\delta}; \\ &\cdots \\ |a_n| &< H_0^{1-\delta} \end{aligned} \tag{2}$$

has nontrivial solutions in integers $a_0, a_1, \ldots, a_n$, provided that $c > \sup |f_1'(x)|$, $x \in [a, b]$.

Suppose that $|F'(\xi)| < H_0^{1-\delta}$. Then, by using the inequalities of system (2), we find that $|F(\xi)| < H^{-\nu_n - \sigma}$ at the point ξ. Since Γ is ν-extremal, ξ belongs to the set $\mathcal{M}(\nu_n + \delta)$ of measure zero. Such a point ξ does not contradict the validity of Lemma 1.

If the inequality $|F'(\xi)| > H_0^{1-\delta}$ is satisfied at ξ, then by using the first inequality in (2) and the mean value theorem, we can establish the existence of a zero α of $F(x)$ such that $\xi \in R(I, H)$ for sufficiently large H. The resulting contradiction establishes Lemma 1.

Every function $F(x)$ has at most a finite number of zeros on $[a, b]$ (see [2]). Then the set S of all possible zeros of the functions $F(x)$ is countable. For a zero α of the function $F(x) = a_0 + a_1 f_1(x) + \cdots + a_n(x)$ we determine the value of the function N:

$$N(\alpha) = H^{1+(1-\varepsilon)(\nu+n^{-1}-1)^{-1}} \tag{3}$$

where $H = \max(|a_0|, \ldots, |a_n|)$. If the same α is a zero of more than one function $F(x)$, then in (3) we are to understand by H the smallest of the individual numbers $\max(|a_0|, \ldots, |a_n|)$.

Lemma 1 and the considerations of §4 of [4] let us prove the following lemma.

LEMMA 2. *The set of real numbers $\xi \in [a, b]$ for which there exists an infinity of zeros α of $F(x)$ that satisfy the condition $|\xi - \alpha| < (N(\alpha))^{-\sigma}$ has Hausdorff dimension at least equal to σ^{-1}.*

From the inequality $|\xi - \alpha| < H^{-w-1}$ it follows that $|F(\xi)| < H^{-w}$. Determine σ from the relation $(N(\alpha)^{-\sigma}) = H^{-w-1}$. Then

$$\sigma = (w + 1)(1 + (1 - \varepsilon)(\nu + 1/2 - 1)^{-1}).$$

Using Lemma 2, we obtain

$$\dim A_n(w) \geq (1 + (1 - \varepsilon)(\nu + n^{-1} - 1)^{-1})(w + 1)^{-1}.$$

Since the preceding discussion holds for every $\varepsilon > 0$, we obtain Theorem 2 by letting $\varepsilon \to 0$.

From the condition $W(F_1'(x), \ldots, F_n'(x)) \neq 0$, there follows the existence of two functions, say $f_1(x)$ and $f_2(x)$, such that $W(f_1'(x), f_2'(x)) \neq 0$ at all points of a certain interval $[a_1, b_1] \subset [a, b]$. Let $A_2'(w)$ be the set of points $x \in [a_1, b_1]$ such that inequality (1) is satisfied infinitely often with $a_3 = a_4 = \cdots = a_n = 0$; that is, $|a_0 + a_1 f_1(x) + f_2(x)| < H^{-w}$.

By a theorem of Schmidt [5] the curve $\Gamma = (f_1(x), f_2(x))$ is 1-extremal; hence, by Corollary 1, $\dim A_2'(w) \geq 3(w + 1)^{-1}$. Since $A_2'(w) \subset A_n(w)$ then $\dim A_2'(w) \leq \dim A_n(w)$ also for $\nu < 3/2 - 1/n$; the inequality $\dim A_n(w) \geq 3(w + 1)^{-1}$ is better than that in Theorem 2. This establishes Corollary 2.

THEOREM 3. *Let $f_1(x)$ and $f_2(x)$ be twice continuously differentiable and $f_1''(x)f_2'(x) - f_1'(x)f_2''(x) \neq 0$ on the interval $[a, b]$. Then*

$$\dim A_2(w) \leq \min\left(\frac{6}{w+1}, \frac{1}{2} + \frac{3}{2(w+1)}\right).$$

PROOF. The first inequality is a corollary of Theorem 1. To obtain the second, we consider the set

$$\tau_F = \{x \in [a, b] \colon |F(x)| \leq H^{-w/2-\varepsilon_1}\},$$

where $\varepsilon_1 > 0$.

We say that $F(x)$ is a function of type I if $\min |F'(x)| \geq H^{(-w-2)/4}$, and of type II if the opposite inequality is satisfied. In each case we use a division of the set of domains τ_F into essential and inessential ones, using the approach of [6], with the modification that in the case of a function of type I we call domains τ_F essential if they are completely disjoint from those similar to them. This modification is required in replacing Lebesgue measure by Hausdorff measure.

In this way we obtain a division of $A_2(w)$ into subsets for each of which we obtain an upper bound and then choose the largest.

The division of $A_2(w)$ into subsets allows us to improve the proof in order to obtain precise values for $\dim A_2(w)$.

REFERENCES

1. V. I. Bernik, *Induced extremal surfaces*, Mat. Sb. **103(145)** (1977), 480–489; English transl. in Math. USSR. Sb. **32** (1977).
2. A. S. Pyartli, *Diophantine approximation on submanifolds of a Euclidean space*, Funktsional. Anal. i Prilozhen. **3** (1969), no. 4, 59–62; English transl. in Functional Anal. Appl. **3** (1969).
3. J. W. S. Cassels, *An introduction to Diophantine approximation*, Cambridge Univ. Press, 1957.
4. A. Baker and Wolfgang M. Schmidt, *Diophantine approximation and Hausdorff dimension*, Proc. London Math. Soc. (3) **21** (1970), 1–11.
5. Wolfgang M. Schmidt, *Metrische Sätze über simultane Approximation abhängiger Grössen*, Monatsh. Math. **68** (1964), 154–166.
6. V. G. Sprindzhuk, *Metric theory of Diophantine approximation*, "Nauka", Moscow, 1977; English transl., Wiley, 1979.

Translated by RALPH P. BOAS

Amer. Math. Soc. Transl.
(2) Vol. **134**, 1987

On the Lattice of Semigroup Varieties

A. YA. AĬZENSHTAT AND B. K. BOGUTA

During the last ten to fifteen years the theory of semigroup varieties has begun developing very fast in various directions. In particular, quite a few works devoted to the properties of the lattice of semigroup varieties have been published. In 1971 the first survey article on various problems of the theory of semigroup varieties was published (Evans [36]). This paper played a significant role in the further development of the theory. During the seven years which have passed since then many new papers on semigroup varieties have been published, including some in which the problems raised in Evans' paper have been solved completely or in part. One chapter in the latest edition of Lyapin's monograph [78] is dedicated to the theory of semigroup varieties; it contains some facts on the lattice of semigroup varieties (with complete proofs).

This paper is also a survey. Our main goal is to survey the results dedicated to one directon in the theory of semigroup varieties—properties of the lattice of the varieties. Also, taking into consideration the role played by the free semigroups of varieties, we have devoted a section to a survey of the papers dedicated to such semigroups. All the results are given without proofs. Of course, some papers are discussed in more detail than others; some are merely mentioned. In particular, the papers considered in Evans' survey [36] are mentioned very briefly. The list of references contains not only the ones mentioned in the present survey but also all papers on the theory of semigroup varieties known to us. However, we do not give references to investigations of properties of semigroups of such well-known classes as the commutative semigroups, the idempotent semigroups, etc.

Let $\mathfrak{X} = \{x_1, x_2, \ldots\}$ be a countably infinite alphabet. Its elements will be called *letters* or *variables*. $\mathfrak{W}(\mathfrak{X})$ denotes the semigroup of words over $\mathfrak{X}$, i.e. it

1980 *Mathematics Subject Classification* (1985 *Revision*). Primary 20-02, 20M07; Secondary 08B15.

Translation of Semigroup Varieties and Semigroups of Endomorphisms (E. S. Lyapin et al., editors), Leningrad Gos. Ped. Inst., Leningrad, 1979, pp. 3–46; MR **81g:** 20108.

is a free semigroup with the free basis $\mathfrak{X}$. If $u \equiv x_{i_1} \cdots x_{i_n}$ is a word over $\mathfrak{X}$, where $\equiv$ denotes the equality of elements of $\mathfrak{W}(\mathfrak{X})$, then $l(u)$ is the length n, and $l_{x_i}(u)$ is the number of occurrences of x_i in u, while $f(u)$ is the number of all variables occurring in u. The elements of $\mathfrak{X}$ will be denoted x, y, z as well.

A pair of words over $\mathfrak{X}$ connected by an equality sign is called a *semigroup identity* or merely an *identity*. If δ is a one-to-one mapping of $\mathfrak{X}$ into itself, then the identities $u = v$ and $\delta u = \delta v$ are usually considered as equivalent. An identity $u = v$ holds in a semigroup A if the equality $\varphi u = \varphi v$ holds for every mapping φ of $\mathfrak{X}$ into A. An identity of the form $u = u$ is called trivial.

An identity $u = v$ is called *normal* (or *regular*, or *homotypical*) if $f(u) = f(v)$; otherwise $u = v$ is called *anomalous*. If $l_x(u) = l_x(v)$ for every variable x in $u = v$, then the identity is called *balanced*. A balanced identity is called a *permutational* one if each of its variables occurs precisely once in each of its words.

For an arbitrary system Φ of identities let $D(\Phi)$ denote the greatest common divisor of all differences $l_x(u) - l_x(v)$, where $u = v$ is an identity from Φ and $x \in \mathfrak{X}$. (This characteristic was considered by Biryukov in [12].)

Let K be a class of semigroups, and let Φ be a set of identities. An identity τ is called a *consequence* of Φ in K if it holds in every semigroup from K in which the identities from Φ hold. A consequence in the class of all semigroups is called a consequence of Φ. Suppose that $u = v$ is an identity from Φ, φ is an endomorphism of $\mathfrak{W}(\mathfrak{X})$, and w_1 and w_2 are words or empty symbols. Then, obviously, the identity $w_1 \varphi u w_2 = w_1 \varphi v w_2$ is a consequence of Φ; it is called an *immediate* consequence of Φ. It is also obvious that if $u_1 = u_2$ and $u_2 = u_3$ are consequences of Φ, then $u_1 = u_3$ is a consequence of Φ. It follows that if the identities $u_1 = u_2$, $u_2 = u_3, \ldots, u_{n-1} = u_n$ are immediate consequences of Φ, then $u_1 = u_n$ is a consequence of Φ. Every nontrivial consequence of Φ can be obtained in this way; this fact is both well known and widely used at present. It seems that this statement was first published by Biryukov in 1963 [12]. Analogous theorems for consequences relative to other interesting classes of semigroups different from the class of all semigroups are not known, and this makes finding the consequences even more difficult.

A system Φ_2 of identities *follows from* a system Φ_1 if every identity from Φ_2 is a consequence of Φ_1. If Φ_2 follows from Φ_1 and vice versa, then the systems Φ_1 and Φ_2 are called *equivalent*. A system of identities is called *irreducible* if no identity of the system follows from the remaining identities. Of course, definitions of a consequence, equivalence, and irreducibility can be given for an arbitrary class of semigroups.

A system Φ_1 of identities which is contained in a system Φ is called a *base* of Φ if every identity of Φ is a consequence of Φ_1.

A class K consisting of all those semigroups in which all the identities of a system Φ hold is called *the variety determined by the system of identities*. This variety is denoted by $\Pi(\Phi)$. Varieties of semigroups are precisely those classes of semigroups which are closed under subsemigroups, homomorphic images, and Cartesian products (cf. [82] and [78]).

The class $\mathfrak{L}$ of all semigroup varieties is a complete lattice under inclusion. If V_α is a class of varieties from $\mathfrak{L}$, then inf V_α is the intersection of all varieties V_α, and is denoted by $\bigwedge_\alpha V_\alpha$. The least variety containing all the varieties from V_α is sup V_α, denoted by $\bigvee_\alpha V_\alpha$.

It is easily seen that if varieties V_α are determined by systems Φ_α of identities, then the variety $\bigwedge_\alpha V_\alpha$ is determined by the system Φ which is the union of all Φ_α, while the set of all identities which hold in $\bigvee_\alpha V_\alpha$ is the intersection of the sets of identities which hold in all varieties V_α. It follows that every variety is an intersection of a set of varieties determined by single identities. This circumstance shows that the latter varieties are an interesting object.

The variety of one-element semigroups is the least element of the lattice $\mathfrak{L}$, while the variety of all semigroups is its greatest element. The atoms of $\mathfrak{L}$ are well known. They are the following varieties:

$$\begin{aligned}
\Gamma_l &= \Pi(xy = x),\\
\Gamma_r &= \Pi(xy = y),\\
\Gamma_o &= \Pi(x_1x_2 = x_3x_4),\\
\Gamma_s &= \Pi(x^2 = x,\ xy = yx),\\
\Gamma_p &= \Pi(x^py = y,\ xy = yx)
\end{aligned}$$

where p is an arbitrary prime integer [62]. The atoms of the form Γ_p are called group atoms.

It is known that $\mathfrak{L}$ has no anti-atoms. It is also known that the variety of all semigroups is not a join of a finite number of proper subvarieties [27].

In this paper we may often consider varieties determined by systems of balanced identities. Since each balanced identity follows from $xy = yx$, these varieties contain the variety $\Pi(xy = yx)$ of all commutative semigroups. We call such varieties *supracommutative*.

All needed definitions and information from the theory of semigroups can be found in the monographs [26] and [77].

§1. The cardinality of $\mathfrak{L}$ and its sublattices. Embedding in irreducible systems

1.1. Let V' be a subvariety of a variety $V = \Pi(\Phi)$. Then V' can be determined by a system of identities containing Φ, i.e. $V' = \Pi(\Phi \cup \Phi')$. The system Φ' can be naturally called a *base of identities of* V' *over* V. When it is clear which V is meant, we will often omit Φ altogether.

1.2. Let $V = \Pi(\Phi)$ be a variety. The lattice of subvarieties of V is denoted by $\mathfrak{L}_V$ or $\mathfrak{L}_\Phi$. Obviously, $\mathfrak{L}_V$ is a complete sublattice of $\mathfrak{L}$. Of course, $\mathfrak{L}$ is $\mathfrak{L}_\Phi$ for the empty set Φ of identities.

1.3. If for every subvariety of $\Pi(\Phi)$ there exists a finite base of identities over $\Pi(\Phi)$, then the lattice $\mathfrak{L}_\Phi$ is at most countably infinite. If Φ is finite, this means that every subvariety of $\Pi(\Phi)$ has a finite base of identities (i.e., it is finitely based). Such a variety is called *hereditarily finitely based.* If $\Pi(\Phi)$ has a subvariety V' without a finite base of identities over $\Pi(\Phi)$, then the cardinality of $\mathfrak{L}_\Phi$ cannot be determined in the general case. In particular cases, when $V' = \Pi(\Phi \cup \Phi')$ and the system $\Phi \cup \Phi'$ is irreducible, the lattice $\mathfrak{L}_\Phi$ has the cardinality of the continuum. Indeed, in this case there exists a continuum of subsets of Φ' which, together with Φ, determine distinct subvarieties of $\Pi(\Phi)$.

1.4. It follows from 1.3 that if Φ is a finite system of identities which is embeddable in an infinite irreducible system of identities, then $\mathfrak{L}_\Phi$ has the cardinality of the continuum. Thus, the existence of infinite irreducible systems of identities suffices for $\mathfrak{L}$ to have the cardinality of the continuum.

1.5. The first systems of identities having no finite base were found independently by Biryukov [11] and Austin [8]. Biryukov proved that a system of identities of the form $u_1vu_2 = u_1u_2$, where $\chi(u_1) = \chi(u_2) \supset \chi(v)$, has no finite base. In the proof the following irreducible system is used:

$$x_1x_2\cdots x_nx_1x_{n-1}\cdots x_1x_n = x_1x_2\cdots x_nx_{n-1}\cdots x_1x_n \qquad (n = 1, 2, \ldots). \quad (1)$$

Austin proved that the system of all balanced identities in which each of the variables occurs at least twice has no finite base. In his proof he used words of the form $x_nx_1\cdots x_{n-1}x_nx_{n-1}\cdots x_1$ $(n = 3, 4, \ldots)$. Subwords of length greater than one are contained in these words precisely once. Evans used an irreducible system

$$x_1\cdots x_nx_1\cdots x_n = x_nx_1\cdots x_{n-1}x_nx_{n-1}\cdots x_1 \qquad (n = 3, 4, \ldots) \quad (2)$$

of identities to prove that $\mathfrak{L}$ has the cardinality of the continuum [32]. In fact, what Evans proved was that if Φ is an infinite irreducible system of identities, then the lattice of varieties containing $\Pi(\Phi)$ has the cardinality of the continuum. In the general case nothing is known about the cardinality of $\mathfrak{L}_\Phi$.

1.6. At present quite a few systems of infinite irreducible systems of identities are known (see [79], [8], [11], [6], and other papers). A majority of such systems consist of balanced identities, and a majority of these latter systems consists of identities in which the number of variables grows indefinitely. The Biryukov system consists of normal but not balanced identities.

1.7. We will note a few other irreducible systems of identities. The system

$$x_1x_2x_1^kx_2 = x_2x_1^kx_2x_1 \qquad (k = 2, 3, \ldots) \quad (3)$$

of balanced identities in two variables is irreducible in the class of all finite semigroups [79]. The system

$$\begin{aligned} \tau_n = (&x_1x_2\cdots x_{n-1}x_nx_2x_nx_{n-1}\cdots x_2x_1x_2 \\ &= x_2x_1x_2\cdots x_nx_2x_nx_{n-1}\cdots x_2x_1) \qquad (n = 4, 5, \ldots) \end{aligned} \quad (4)$$

of balanced identities is irreducible in the class of semigroups having no idempotents [80]. Of course, if a system of identities is irreducible in some class of semigroups, it is irreducible in the class of all semigroups.

1.8. Let Φ consist of a single identity $u = v$. When does $\mathfrak{L}_\Phi$ have the cardinality of the continuum? In particular, when is it possible to include $u = v$

in an infinite irreducible system of identities? It turned out that this is "almost always" possible. The first to attack this problem for a balanced identity $u = v$ ($u \not\equiv v$) was Lyapin [39]. He found that if, in a balanced identity $u = v$, both u and v contain squares of variables, or if this identity contains n variables while the lengths of both u and v are greater than $2n+2$, then there exists a positive integer s such that the system of identities $u = v,\ \tau_s,\ \tau_{s+1}, \ldots$ is irreducible in the class of semigroups having no idempotents. Here the τ_i are the identities from (4). It follows that every balanced identity over the alphabet $\{x_1, \ldots, x_n\}$ ($n = 1, 2, \ldots$) with the exception of finitely many can be embedded in an infinite system of identities which is irreducible in the class of semigroups having no idempotents.

1.9. What can we say about the identities not listed in 1.8 or the unbalanced identities? These identities were considered by Pollák in [115] and [116].

Let Σ denote a system consisting of the identities of the form

$$x_1 \cdots x_n = v, \quad x_1x_2x_1 = v, \quad x_1x_2x_3x_1 = x_i^2,$$

where v is an arbitrary word and x_i is an arbitrary variable, and of identities of the form

$$x_1 \cdots x_n xyx y_1 \cdots y_k = w,$$

where one of the words $x_1 \cdots x_n$ and $y_1 \cdots y_k$ may be empty and where $w = u_1 \cdots u_m$. Here the words $u_1, \ldots, u_m$ must consist of differing variables and u_i always is either a square of a variable or a word in which each of its variables occurs only once. An arbitrary identity which does not belong to Σ can be embedded in an infinite irreducible system of identities. As a corollary we obtain that if a balanced identity $u = v$ is not a permutational one and is not an identity of the form

$$x_1 \cdots x_{i-1} y x_i y x_{i+1} \cdots x_n = x_{\pi 1} \cdots x_{\pi(j-1)} y^2 x_{\pi j} \cdots x_{\pi n}$$

where π is a permutation of the set $\{1, 2, \ldots, n\}$, then it is embeddable in an infinite irreducible system of identities. In the proof eight infinite irreducible systems are used. One of them is the system containing the words introduced by Austin, and the other ones are found by Pollák, for example, the system

$$yx_1^2 \cdots x_n^2 y = yx_n^2 \cdots x_1^2 y \qquad (n = 2, 3, \ldots). \tag{5}$$

1.10. Which of the identities belonging to Σ determine hereditarily finitely based varieties? This is known for some special types of the identity $x_1 \cdots x_n = v$. For example, the variety $\Pi(xy = yx)$ of commutative semigroups is hereditarily finitely based (Perkins [110]). The Perkins proof is given also in [36]. A variety which is defined by a single permutational identity in which both words either begin or end with different variables is hereditarily finitely based [5]. The following fact is essentially used in the proof. Let τ be a permutational identity of the form

$$y_1 \cdots y_p x_1 \cdots x_n z_1 \cdots z_q = y_1 \cdots y_p x_{\sigma 1} \cdots x_{\sigma n} z_1 \cdots z_q \qquad (\sigma 1 \neq 1,\ \sigma n \neq n),$$

then there exists a positive integer k such that every permutational identity

$$y_1 \cdots y_p x_1 \cdots x_l z_1 \cdots z_q = y_1 \cdots y_p x_{\delta 1} \cdots x_{\delta l} z_1 \cdots z_q$$

($\mathfrak{L} > k$ and δ is a permutation) is a consequence of τ [120].

1.11. The variety determined by a permutational identity of the form $x_1 \cdots x_n = x_{\sigma 1} \cdots x_{\sigma n}$ ($\sigma 1 = 1$, $\sigma n = n$) is not hereditarily finitely based. Indeed, the system of identities

$$x_1 x_2 x_3 x_4 = x_1 x_3 x_2 x_4, \qquad x_2 x_1^k x_2 = x_1 x_2 x_1^{k-2} x_2 x_1$$

for all positive integers $k > 1$ has no finite base [110]. Obviously, the following irreducible subsystem can be found in our system:

$$x_1 x_2 x_3 x_4 = x_1 x_3 x_2 x_4, \qquad x_2 x_1^p x_2 = x_1 x_2 x_1^{p-2} x_2 x_1 \tag{6}$$

where p is an arbitrary odd prime. If all p are taken to be greater than n, the system

$$x_1 \cdots x_n = x_{\sigma 1} \cdots x_{\sigma n}, \qquad x_2 x_1^p x_2 = x_1 x_2 x_1^{p-2} x_2 x_1$$

is also irreducible.

It is easy to see that if every variable occurs only once in v and the identity $x_1 \cdots x_n = v$ is not permutational, then the lattice of subvarieties of

$$\Pi(x_1 \cdots x_n = v)$$

is finite. Thus, one can determine all hereditarily finitely based varieties defined by a single identity of this form. For an arbitrary word v it is rather hopeless to expect a definite answer at the present time.

1.12. It is known that if a variety is determined by two identities of which one is permutational and the other is unbalanced, then it is hereditarily finitely based. It seems this is all that is known about embedding a system containing more than one identity in an infinite irreducible system of identities.

1.13. The variety $\Pi(x_1 x_2 x_3 x_4 = x_i^2)$ is hereditarily finitely based for all i (Pollák [117]). In the same paper Pollák notices that if the system of identities has the latter of the two forms mentioned in 1.9, and also for arbitrary balanced identities, the problem can be completely solved. Of course, among the identities of the form $x_1 x_2 x_1 = v$ there are the identities determining hereditarily finitely based varieties. As an example we can consider $x_1 x_2 x_1 = x_3 x_4$ (commutativity follows from it). However, at present there is no hope for a complete solution to this problem.

1.14. The irreducible systems of identities which we have mentioned do not contain any anomalous identities. However, a slight modification of these systems produces a system with anomalous identities. For example, the system

$$x_1 \cdots x_n y^2 x_1 \cdots x_n = x_n x_1 \cdots x_{n-1} x_n x_{n-1} \cdots x_1 \qquad (n = 3, 4, \ldots) \tag{7}$$

is irreducible. If s is an integer greater than one, then the system consisting of the identity $x^s = x^{s+1}$ and the identities from (7) in which $n > s$ is irreducible [6]. This system determines a variety containing a single atom $\Pi(x_1 x_2 = x_3 x_4)$;

hence it consists of nilsemigroups only. Varieties consisting only of nilsemigroups form a continuous sublattice of $\mathfrak{L}$. Moreover, even the lattice of the varieties of nilsemigroups of degree not greater than s ($s > 1$) has the cardinality of the continuum [6]. Lysenko [81] found an infinite irreducible system of anomalous identities and used it to prove that the interval $[V_{n-1}; V_n]$ has the cardinality of the continuum. Here V_k ($k = n-1, n$) is the variety of all nilsemigroups of degree not greater than k.

1.15. Suppose that A is a finite semigroup and $\operatorname{Var} A$ is the variety generated by A, i.e. the variety determined by the system of all identities which hold in A. It is known that $\operatorname{Var} A$ need not be finitely based. Perkins [110] found a six-element inverse semigroup with zero and identity without any finite base of identities. Is this semigroup minimal? It is known that the system of all identities of an arbitrary two-element algebra is finitely based. For a three-element algebra this is not always true (see Evans [36]). However, a three-element semigroup is always finitely based [110]. The same holds for every four-element semigroup [19], [139]. Since every commutative and every idempotent semigroup is finitely based, one needs to consider only nonidempotent and noncommutative four- and five-element semigroups. It turns out that every nonidempotent and noncommutative four-element semigroup has a base of identities consisting of no more than four identities, and these identities contain no more than three variables [139]. Also, finite bases of identities can be found for all noncommutative monoids with zero containing five elements at most (there are 19 five-element ones) [29]. Thus, among the monoids with zero the Perkins semigroup is minimal having no finite base of identities. If it is minimal in the class of all semigroups is as yet unknown.(1)

1.16. The following problem is interesting. Can every variety of semigroups be determined by an irreducible system of identities? Every system of balanced identities contains an equivalent irreducible subsystem [3]. Therefore every supracommutative variety can be determined by an irreducible system of identities. For arbitrary varieties this does not hold. Trakhtman [153] gave the following example of a variety which possesses no irreducible base of identities:

$$\Pi(yx^4y = yx^5y,\ yxyz_1^2 \cdots z_n^2x^6y = y^2xz_1^2 \cdots z_n^2x^6y) \qquad (n = 1, 2, \ldots).$$

This variety does not contain any group atoms, i.e. it has no nontrivial groups; however, it contains all other atoms.

1.17. Let a variety be given. The following problem is quite natural: does there exist, for every finitely based subvariety of this variety, a base of identities such that the number of identities in each of these bases, or the number of variables occurring in all these bases, is bounded? An analogous question is: can one embed a given identity in an irreducible system with any finite number of identities? In certain cases answers are known. For example, it is well known that every idempotent variety can be determined by two identities (see, for example,

(1) *Translator's note.* It was announced in [154] that every five-element semigroup is finitely based. No indication of the method of proof was given.

[15]). The number of variables appearing in the bases of identities of varieties of commutative semigroups is not bounded from above [110]. Every balanced identity can be embedded in an irreducible system of identities with any finite number of identities [107].

1.18. The following question is also interesting: for which varieties V is the lattice $\mathfrak{L}_V$ finite? This problem was raised in Evans' survey [36] (Evans called such varieties *small*). If $V = \Pi(\Phi)$ and Φ does not contain an unbalanced identity of the form $x_1 \cdots x_n = v$, then $\mathfrak{L}_V$ is infinite, because it contains the infinite chain

$$\Pi(\Phi, x_1x_2 = x_1x_2^2) \subset \Pi(\Phi, x_1x_2x_3 = x_1x_2x_3^2) \subset \cdots \subset \Pi(\Phi).$$

On the other hand, the example of the variety of idempotent semigroups shows that if an unbalanced identity of the form $x_1 \cdots x_n = v$ belongs to Φ, then the lattice $\mathfrak{L}_\Phi$ need not be finite.

1.19. If Φ contains both the identity of commutativity and an unbalanced identity of the form $x_1 \cdots x_n = v$, then the lattice of subvarieties of $\Pi(\Phi)$ is finite [7]. Therefore, all commutative varieties V with finite $\mathfrak{L}_V$ are known, because if a base of identities of a variety V contains an unbalanced identity of the form $x_1 \cdots x_n = v$, then any other base has the same property.

1.20. Note that (see [7]) every commutative variety which satisfies an unbalanced identity $x_1 \cdots x_n = v$ is a subvariety of the variety

$$\Pi(x_1x_2 = x_2x_1,\ x_1 \cdots x_{n-1}x_n = x_1 \cdots x_{n-1}x_n^k).$$

Therefore, it is interesting to study the lattices of subvarieties of such varieties (see §4).

1.21. The results of 1.19 and 1.20 hold for those varieties $\Pi(\Phi)$ for which Φ contains a permutational identity of the form $x_1 \cdots x_n = x_{\sigma 1} \cdots x_{\sigma n}$, where σ is a permutation of $\{1, \dots, n\}$ and either $\sigma 1 \neq 1$ or $\sigma n \neq n$. Note that even if σ is an arbitrary permutation and Φ contains an unbalanced identity $x_1 \cdots x_n = v$, then $\Pi(\Phi)$ is a subvariety of a variety of the form (see [7])

$$\Pi(x_1 \cdots x_n = x_{\sigma 1} \cdots x_{\sigma n},\ x_1 \cdots x_m = x_1 \cdots x_i^k \cdots x_m).$$

It is not known if this latter variety always has finitely many subvarieties.

1.22. Some information on small varieties can be found in [6] and [9].

§2. Coverings in the lattice of semigroup varieties

2.1. Does every variety of algebras have a covering element in the lattice of all varieties? For semigroups this problem was mentioned by Evans in [36]. It has been considered for other algebras as well, e.g., for lattices.

The first advance in solving this problem for the lattice of semigroup varieties was made in [3]. It was proved that every supracommutative semigroup variety has a cover in the lattice of all semigroup varieties. Trakhtman [151] proved the following theorem.

Let U be a variety of algebras determined by balanced identities of type Ω which does not contain unary operations. Then every variety from $\mathfrak{L}_U$ but U has a covering variety in $\mathfrak{L}_U$.

In particular, this theorem gives a positive solution to the problem of Evans.

2.2. It is known (cf. [40]) that the lattice $\mathfrak{L}$ of all semigroup varieties is anti-isomorphic to the lattice of all fully invariant congruences on the semigroup $\mathfrak{W}(\mathfrak{X})$. Here a congruence ρ is called *fully invariant* if, for every endomorphism φ of $\mathfrak{W}(\mathfrak{X})$, $(a,b) \in \rho$ implies $(\varphi a, \varphi b) \in \rho$. The anti-isomorphism is established in the following way: a fully invariant congruence ρ_V corresponds to a variety V if any identity $a = b$ holds in V precisely when $(a,b) \in \rho_V$. The existence of this anti-isomorphism makes it possible to reduce the problem of the existence of covers for a variety V to the problem of the existence of maximal fully invariant congruences of $\mathfrak{W}(\mathfrak{X})$ which are contained in ρ_V, where ρ_V is the fully invariant congruence corresponding to V. The proof of the theorem stated in 2.1 exploits this idea.

2.3. The theorem in 2.1 gives a wide class of varieties V such that every variety but V in the lattice $\mathfrak{L}_V$ has a cover. However, not every variety has this property. For example, the variety $V = \Pi(yx^4y = yx^5y)$ contains a subvariety

$$U = \Pi(yx^4y = yx^5y,\ yxyz_1^2 \cdots z_n^2x^6y = y^2xz_1^2 \cdots z_n^2x^6y,\ n = 1, 2, \ldots)$$

which has no covers in $\mathfrak{L}_V$ [153]. As we mentioned in 1.16, this variety cannot be determined by any irreducible system of identities.

2.4. Since every variety of semigroups has a cover, one is naturally interested in the cardinality of the set of all covers of a given variety and in the dependence of this cardinality on the minimum cardinality of a system of identities determining the variety. Possible answers vary very much. For example, the variety of commutative semigroups $\Pi(xy = yx)$ has a single cover $\Pi(xyz = yxz,\ xyz = xzy)$ (see [36]). The variety of trivial semigroups $\Pi(x = y)$ is an example of a variety which is determined by a single identity and which has an infinite countable number of covers [62].

The variety U which was considered in 2.3 is defined by a countably infinite system of identities without an irreducible base, and it has a single cover in the lattice of all semigroup varieties.

2.5. There is an example of a variety which is determined by an infinite irreducible system of identities and which has continuum many covers [151]. This variety is constructed as follows. Consider an infinite sequence of words of the form xy, $xyyx$, $xyyxyxxy$, and so on, each of the words being obtained from the previous one by substituting xy for x and yx for y. Next, in each of the words of the sequence which have x as their last letter, strike out this occurrence of x, and add y^4 at the beginning and x^4 at the end of the resulting word. Denote by $u_1, u_2, \ldots, u_n, \ldots$ the sequence so obtained. The system of identities $u_1 = u_2,\ u_1 = u_3, \ldots, u_1 = u_n, \ldots$ is irreducible, and the variety $\Pi(u_1 = u_2,\ u_1 = u_3, \ldots, u_1 = u_n, \ldots)$ has a continuum of covers in the lattice of all varieties.

Though the author of [151] does not mention it, in fact he has constructed a continuum of varieties each of which has a continuum of covers, since every infinite subsystem of his system of identities determines a variety having a continuum of covers.

2.6. The problem of covers in supracommutative varieties has been studied in more detail. Every supracommutative variety can be determined by an irreducible system of identities; and if this system is finite, then the variety has finitely many covers; otherwise it has a countably infinite set of covers [3].

Estimates for the number of covers of finitely based supracommutative varieties have been found. In certain cases the identities determining a cover can be found from the identities determining the original variety (see [18]).

2.7. The question of the number of covers for an arbitrary variety determined by a single identity was studied in [17]. The following facts were established there. Every variety which is determined by a single anomalous identity has infinitely many covers. Every variety determined by a single identity of the form $u = v_1 u v_2$, where u, v_1, and v_2 are arbitrary words (one of the words v_1 or v_2 may be empty), has infinitely many covers. A variety has finitely many covers if it is determined by a single normal identity $u = v$ which satisfies one of the following two conditions: (1) $\mathfrak{L}(u) = l(v)$; or (2) $\mathfrak{L}(u) < l(v)$, but $v \not\equiv v_1 \varphi u v_2$ for all (possibly empty) words v_1 and v_2 and every mapping $\varphi\colon \mathfrak{X} \to \mathfrak{W}(\mathfrak{X})$.

2.8. The proof of the statements given in 2.7 is carried out in the language of congruences. However, it is easily seen that the covers constructed in the proof are joins of the varieties and group atoms $\Pi(xy = yx, x^p y = y)$. This gives another approach to constructing covers of these varieties.

2.9. Let R be an arbitrary but fixed variety of semigroups. A semigroup A is called an *mR-semigroup* if all proper subsemigroups of A belong to R, while A does not belong to R. A minimal variety containing R and a fixed mR-semigroup is called an *mR-variety*. Let $V_1 R$ denote the minimal variety containing R and all mR-semigroups which are not groups. Let $V_2 R$ denote the minimal variety containing R and all mR-varieties.

The problem of the interrelations between the mR-varieties $V_1 R$, $V_2 R$, and R in the lattice of all semigroups was considered in [152]. The intervals $[R; V_1 R]$ and $[R; R_1]$, where R_1 is an mR-variety, may be sufficiently large. There was given an example of a variety R and an mR-semigroup such that these intervals do not satisfy either the minimality or the maximality condition.

However, if R is an arbitrary variety of idempotent semigroups, then the set of mR-varieties coincides with the set of covers of R. As a corollary of this fact, descriptions of the corresponding varieties $V_1 R$ and $V_2 R$ were found.

2.10. The covers of the varieties $Z_n = \Pi(x_1 \cdots x_n = y_1 \cdots y_n)$ of all n-nilpotent semigroups have been completely described [2]. There are countably infinitely many of them, and each one of them may be determined by at most four identities. Only one of the covers

$$V_n^{(0)} = \Pi(x_1 \cdots x_n = x_2x_1 \cdots x_n,\ x_1 \cdots x_n = x_2x_3 \cdots x_nx_1,$$
$$x_1 \cdots x_nx_{n+1} = y_1^2y_2 \cdots y_{n-1})$$

consists of nilpotent semigroups. All other covers are joins of Z_n and an atom.

It was established in [2] that an arbitrary variety V_n of n-nilpotent semigroups has at least two covering varieties, and if V_n can be determined by a system of identities in which every nontrivial identity has the form $x_1x_2 \cdots x_n = v$, then there exists a countably infinite set of covers; otherwise there are only finitely many covers.

2.11. In §4 we will mention covers for certain concrete classes of commutative varieties.

§3. On sublattices of $\mathfrak{L}$

3.1. We have already mentioned that the lattice $\mathfrak{L}$ has the cardinality of the continuum. A complete description of $\mathfrak{L}$ in any sense whatever is beyond any hope at this time. However, quite a few facts are known.

3.2. Here we consider various sublattices of $\mathfrak{L}$, in particular, finite ones. If V is a variety, then, as we have already mentioned, the set $\mathfrak{L}_V$ of all its subvarieties is a complete sublattice of $\mathfrak{L}$. The lattices of varieties of commutative semigroups and of idempotent semigroups are countably infinite lattices of this type. A complete description of the latter lattice is known (see [15], [39], and [45]), and a good deal is known about the lattice of commutative semigroup varieties (see [4]).

3.3. Using his description of the lattice $\mathfrak{L}_{(x^2=x)}$, Gerhard [50] investigated the lattice $\mathfrak{L}_2$ of subvarieties of the variety $\Pi((xy)^2 = xy)$. He defined a congruence θ on $\mathfrak{L}_2$ in the following natural way: $\Pi(\Phi_1)\ \theta\ \Pi(\Phi_2)$ if the varieties $\Pi(x^2 = x, \Phi_1)$ and $\Pi(x^2 = x, \Phi_2)$ coincide. Clearly, the lattice $\mathfrak{L}_2/\theta$ is isomorphic to $\mathfrak{L}_{(x^2=x)}$. Since a description of this lattice is available, we may get a description of $\mathfrak{L}_2$ if we know the structure of congruence classes modulo θ. It is easily seen that the class $\theta\Pi(x^2 = x)$ is a dual ideal of $\mathfrak{L}_2$. The lattice $\theta\Pi(x^2 = x)$ has a very simple structure; its description is found. The greatest elements of the remaining classes are found as well, and it is proved that every element of $\mathfrak{L}_2$ can be represented in the form $\Pi(p = q, f = g)$, where $\Pi(p = q)$ is the greatest element of a corresponding class, while $\Pi(f = g)$ belongs to $\theta\Pi(x^2 = x)$. For every class different from $\theta\Pi(x^2 = x)$ there is found a lattice isomorphic ideal of $\theta\Pi(x^2 = x)$. It follows that $\mathfrak{L}_2$ is a subdirect product of $\mathfrak{L}_{(x^2=x)}$ and $\theta\Pi(x^2 = x)$.

3.4. Certain sublattices of $\mathfrak{L}_2$ (cf. 3.3) have been investigated earlier. Mel′nik was interested in the lattice of subvarieties of the variety $B = \Pi(xyzxy = xy)$ because of Schein's problem of describing the semigroups in which the product of any three elements xyz equals one of xy, xz, or yz [94]. $\mathfrak{L}_B$ is a sublattice of $\mathfrak{L}_2$, since the identity $xyzxy = xy$ implies $(xy)^2 = xy$. The lattice $\mathfrak{L}_B$ is distributive; it consists of 13 elements. B consists of semigroups A such that

A^2 is the semigroup of all idempotents of A; it is a rectangular band of trivial groups.(2)

3.5. A natural generalization of B was considered by Petrich [111]. Let $G_n = \Pi(x^n y = yx^n = y)$ be the variety of n-groups. Let C_n be the class of semigroups A such that A^2 is an orthodox normal band of groups from G_n (i.e. A^2 admits a congruence ρ such that every ρ-class is an n-group, the quotient semigroup A^2/ρ is idempotent and it satisfies the identity $xyzx = xzyx$, and the set of idempotents of A^2 is closed under multiplication). C_n is a finitely based variety, namely,

$$C_n = \Pi((xy)^{n+1} = xy,\ x_1x_2x_3^n x_4^n x_5 x_6 = x_1x_2x_4^n x_3^n x_5 x_6).$$

B is a subvariety of this variety, since $xyzx = xzyx$ implies the identities

$$(xy)^2 = xy, \qquad x_1x_2x_3x_4x_5 = x_1x_2x_4x_5.$$

Of course, G_n is a subvariety of C_n as well.

3.6. C_n is the join of subvarieties B, G_n, and Γ_s, where

$$\Gamma_s = \Pi(xy = yx,\ x^2 = x).$$

The lattice $\mathfrak{L}_{C_n}$ possesses the following interesting property: for any varieties $V' \subset B$, $V'' \subset G_n$, and $V''' \subset \Gamma_s$, the lattice $\mathfrak{L}_{V' \vee V'' \vee V'''}$ is isomorphic to the direct product $\mathfrak{L}_{V'} \times \mathfrak{L}_{V''} \times \mathfrak{L}_{V'''}$. Therefore, $\mathfrak{L}_{C_n}$ itself is isomorphic to the direct product $\mathfrak{L}_B \times \mathfrak{L}_{G_n} \times \mathfrak{L}_{\Gamma_s}$. Unfortunately, almost nothing is known about the lattice of subvarieties of G_n.

3.7. It is interesting that the intervals

$$[G_n; B \vee G], \qquad [G_n \vee \Gamma_s;\ B \vee G_n \vee \Gamma_s], \qquad [A_n \vee \Gamma_s;\ B \vee A_n \vee \Gamma_s],$$

where A_n is the variety of abelian n-groups, are isomorphic to the lattice $\mathfrak{L}_B$. All subvarieties of these intervals are found; each of them can be determined by at most two identities [111].

3.8. Quite a few finite sublattices of $\mathfrak{L}$ have been described. We have discussed one such sublattice in 3.4. Some examples can be found in the paper of Evans [36]. In particular, one of the lattices discussed there shows that $\mathfrak{L}$ is not modular (see [60]).

3.9. Let $\mathfrak{L}^{(\mathrm{N})}$ be the lattice of varieties of nilsemigroups, i.e., of the varieties containing $\Pi(x_1x_2 = x_3x_4)$ as their only atom. Let $V \in \mathfrak{L}^{(\mathrm{N})}$. The lattice $\mathfrak{L}_V$ is finite if and only if V is a variety of nilpotent semigroups, i.e. when it is a subvariety of the variety $Z_n = \Pi(x_1 \cdots x_n = y_1 \cdots y_n)$ for some n [6].

3.10. Mel′nik [95] proved that the lattice of subvarieties of Z_n is distributive if and only if $n \le 3$, and the lattice Z_4 is not even modular. Clearly, the only atom of Z_n is the variety $\Pi(x_1x_2 = x_3x_4)$. Its anti-atoms are the varieties

$$\Pi(x_1 \cdots x_n = y_1 \cdots y_n,\ x^{n-1} = x^n),$$

$$\Pi(x_1 \cdots x_n = y_1 \cdots y_n,\ u = v),$$

(2) *Translator's note.* It is easy to see that a semigroup satisfies the identity $(xy)^2 = xy$ precisely when it is an inflation of an idempotent semigroup.

where $l(u) = l(v) = n - 1$, and $u = v$ is a normal identity in two variables. The lattice $\mathfrak{L}_{Z_3}$ is described; it contains precisely six elements. In the same paper analogous problems are solved for the lattice of subvarieties of the variety of n-nilpotent commutative semigroups, and certain sublattices of $\mathfrak{L}$ consisting of commutative semigroups are described.

3.11. Sublattices of $\mathfrak{L}$ can be naturally identified by many other means. For example, let Ω be a set of atoms of $\mathfrak{L}$, let $\mathfrak{L}_\Omega$ be the set of varieties containing each of the atoms from Ω, and let $\mathfrak{L}'_\Omega$ be the set of varieties containing no atoms from Ω. Clearly, $\mathfrak{L}_\Omega$ is a complete sublattice of $\mathfrak{L}$. It turns out that $\mathfrak{L}'_\Omega$ is an ideal of $\mathfrak{L}$; however, it need not be a complete sublattice of $\mathfrak{L}$. Varying Ω we may obtain the lattice $\mathfrak{L}'_\Omega$ consisting of all varieties of regular semigroups, groups, nilsemigroups, etc. [4]. The lattices $\mathfrak{L}_\Omega$ are known when Ω consists of a single atom [141].

3.12. All varieties closed under the standard wreath product (the same varieties are closed under certain other types of wreath products) form a countably infinite distributive complete sublattice of $\mathfrak{L}$. This sublattice has been described. Such varieties are

$$\Pi(x_1x_2\cdots x_k = x_1x_2\cdots x_kx_{k+1}),$$
$$\Pi(x_1x_2\cdots x_k = x_1\cdots x_kx_{k+1}\cdots x_{2k-n}),$$

where $n < k$, $n, k \in \{0, 1, 2, \dots\}$, and, of course, the variety of all semigroups [74].

3.13. Let $V = \Pi(\Phi)$ be a semigroup variety. It is clear that if Φ consists of normal identities, then all identities which hold in V are normal. Such a variety is called *normal.* A variety is called *anomalous* if it cannot be determined by a system of normal identities. For example, every variety consisting only of groups is anomalous.

Let $I(V)$ be the least normal variety containing a variety V; it is called the *normal closure* of V. For any varieties V_1 and V_2 it follows from $I(V_1) \subset I(V_2)$ that $V_1 \subset V_2$ (see [121]). Thus, the equalities $V_1 = V_2$ and $I(V_1) = I(V_2)$ are equivalent for anomalous varieties. Therefore, the normal closure covers an anomalous variety. There exist results concerning the structure of semigroups in the normal closure; we do not consider them here (see [62]).

3.14. Obviously, a variety is normal if and only if it contains the atom $\Gamma_s = \Pi(x^2 = x,\ xy = yx)$. It follows immediately that $I(V) = V \vee \Gamma_s$.

3.15. Suppose we know a base of identities of a variety V. How can we find a base of identities of its normal closure? Such bases are known for many varieties. They can easily be found for atoms, and they have also been found for the varieties of nilpotent semigroups, of rectangular semigroups ($\Pi(xyx = x)$), for the varieties $\Pi(x^n = y^n)$ and $\Pi(x^ny = y)$, and for certain other varieties [92], [121]. Mel′nik in [93] found a general method of constructing a base of identities for a normal closure of a given anomalous variety. Namely, every anomalous variety has a base of identities Φ consisting of a system Φ_1 of normal identities and a single anomalous identity of a special form. A system of identities which is obtained after eight normal identities are added to Φ_1 is a base for $I(V)$. Therefore, if V is finitely based, so is $I(V)$, and vice versa. However, the base of identities of $I(V)$ constructed following this method is not, as a rule, a minimal one.

3.16. The mapping I of $\mathfrak{L}$ into itself which assigns the normal closure $I(V)$ to every variety V is an endomorphism of $\mathfrak{L}$. This mapping I is called the *operator of normal closure* [93]. Such an operator can be naturally considered on the lattice of subvarieties of arbitrary Ω-algebras. It is not always an endomorphism [98].

3.17. We mention another interesting property of the operator of normal closure. For an arbitrary anomalous variety V the lattice of subvarieties of $I(V)$ is isomorphic to a direct product of the lattice $\mathfrak{L}_V$ of subvarieties of V and a two-element chain, i.e. it is a reduplication of $\mathfrak{L}_V$ [93]. Therefore, if we know the structure of $\mathfrak{L}_V$, we also know the structure of $\mathfrak{L}_{I(V)}$.

3.18. Analogously to I, one can consider on $\mathfrak{L}$ an operator (translation) μ_V defined by an arbitrary variety V; it transforms any variety $\overline{V}$ into $\overline{V} \vee V$. The operator I is the translation μ_V with $V = \Gamma_s$. For an arbitrary translation μ_V one can consider the same problems as those considered for I. In particular, when is $\mathfrak{L}_{\mu_V(\overline{V})}$ a reduplication of $\mathfrak{L}_{\overline{V}}$ for every variety $\overline{V}$, or at least for some varieties? It is easy to see that it is necessary that V be an atom.

3.19. Consider the case when the variety V is the atom

$$\Gamma_0 = \Pi(x_1x_2 = x_3x_4).$$

The translation $\mu_V = \mu$ is called *nilpotent.* This translation is a one-to-one endomorphism(3) on $\mathfrak{L}$. For arbitrary variety $\overline{V}$ the lattice $\mathfrak{L}_{\mu(\overline{V})}$ is a reduplication of $\mathfrak{L}_{\overline{V}}$. Mel'nik [96] found a general method describing a base of identities of $\mu(\overline{V})$ when a base of identities for $\overline{V}$ is known.

3.20. Translations determined by other atoms as well as translations satisfying certain special conditions have been considered. As applications, certain new sublattices of $\mathfrak{L}$ were described in [97] and [100].

§4. The lattice of commutative semigroup varieties

4.1. We mentioned in 1.10 that every commutative semigroup variety is finitely based; hence the lattice $\mathfrak{L}_C$ of all commutative semigroup varieties is countably infinite [110].

4.2. The lattice $\mathfrak{L}_C$ is not modular. An example is given in [124]. Let

$$V_1 = \Pi(xy = yx,\ xy^9 = x^2y^8,\ x^3y^7 = x^4y^6),$$
$$V_2 = \Pi(xy = yx,\ x^2y^8 = x^3y^7,\ x^4y^6 = x^5y^5),$$
$$V_3 = \Pi(xy = yx,\ xy^9 = x^2y^8,\ x^3y^7 = x^4y^6,\ xy^9 = x^5y^5).$$

---

(3) *Translator's note.* This statement is trivially false because $\overline{V}$ and $\overline{V} \vee \Gamma_0$ always have the same image under μ. There is no such statement in [96].

Then $V_1 \supset V_3$, but $V_1 \wedge (V_2 \vee V_3) \neq (V_1 \wedge V_2) \vee V_3$ (since the right-hand side is a variety satisfying the identity $xy^9 = x^5y^5$, while the left-hand side does not satisfy it).

Later Burris and Nelson [21] proved that $\mathfrak{L}_C$ does not satisfy any nontrivial lattice identity.

4.3. A large distributive sublattice of $\mathfrak{L}_C$ was pointed out in [125]. It consists of those commutative varieties which can be determined by identities of the form $u = ux^n$, where u is a word and x is a variable. This result was generalized by Nelson [106]. She found a distributive sublattice which is a maximal modular sublattice of $\mathfrak{L}_C$. She also showed that not every modular sublattice of $\mathfrak{L}_C$ is distributive.

4.4. In [136] there is a description of the lattice of commutative varieties $A_{m,n}$, where $A_{0,n}$ is the variety of groups of exponent n, $A_{0,1}$ is the trivial variety, and $A_{m,n}$ is the variety all of whose identities are of the form $x_1^{\alpha_1}x_2^{\alpha_2}\cdots = x_1^{\beta_1}x_2^{\beta_2}\cdots$, where α_i, $\beta_i \geq 0$, $\alpha_i \equiv \beta_i \pmod n$, and $\alpha_i = \beta_i$ whenever α_i or β_i does not exceed m. It is easy to check that

$$A_{k,l} \vee A_{m,n} = A_{\max(k,m),\mathrm{LCM}(l,n)}$$

and

$$A_{k,l} \wedge A_{m,n} = A_{\min(k,m),\mathrm{GCD}(l,n)}.$$

This lattice is isomorphic to a direct product of the lattice of nonnegative integers with the usual order relation and the lattice of division-ordered positive integers.

The intervals $[A_{k,l}; A_{m,n}]$ and lattices $\mathfrak{L}_{A_{m,n}}$ were studied in [105].

4.5. The most detailed information on the structure of the lattice is given in the paper [106] already mentioned. Here are the main results of that paper. Suppose we are given a nontrivial identity of the form

$$u \equiv x_1^{\alpha_1}x_2^{\alpha_2}\cdots x_n^{\alpha_n} = x_1^{\beta_1}x_2^{\beta_2}\cdots x_n^{\beta_n} \equiv v, \qquad \alpha_i, \beta_i \geq 0$$

(since we consider commutative varieties only, every identity can be written in this form). Consider the following characteristics of the identities:

$$\begin{aligned} D(u=v) &= \mathrm{GCD}\{|\alpha_i - \beta_i|\},\\ V(u=v) &= \min\{\alpha_i, \beta_i\},\\ L(u=v) &= \min\{\max\alpha_i,\ \max\beta_i\},\\ U(u=v) &= \begin{cases} \min\{\sum_i \alpha_i, \sum_i \beta_i\} & \text{if } \sum_i \alpha_i \neq \sum_i \beta_i,\\ \sum_i \alpha_i + V(u=v) & \text{if } \sum_i \alpha_i = \sum_i \beta_i. \end{cases} \end{aligned}$$

Analogously, these characteristics are defined for systems of identities.

If $u = v$ is a consequence of a system Φ of identities, then $U(\Phi) \leq U(u = v)$, $V(\Phi) \leq V(u = v)$, $\mathfrak{L}(\Phi) \leq L(u = v)$, and $D(\Phi)$ divides $D(u = v)$.

Let r, s, and n be nonnegative integers with $r \leq s$ and $n \geq 1$. Consider the commutative varieties

$$\Omega_{r,s,n} = \Pi(xy = yx,\ x^r y^s = x^{r+n}y^s,\ x^s = x^{s+n}).$$

An identity $u = v$ holds in $\Omega_{r,s,n}$ if and only if $r \leq V(u = v)$, $s \leq L(u = v)$, and n divides $D(u = v)$. It follows that $\Omega_{r,s,n} \vee \Omega_{t,u,m} = \Omega_{v,w,p}$, where $v = \max\{r, t\}$, $w = \max\{s, u\}$, $p = \mathrm{LCM}\{n, m\}$, and $\Omega_{r,s,n} \wedge \Omega_{t,u,m} = \Omega_{v,w,d}$, where $v = \min\{r, t\}$, $w = \min\{s, u\}$, and $d = \mathrm{GCD}\{n, m\}$.

Thus, the varieties $\Omega_{r,s,n}$ form a sublattice which Nelson calls a "skeleton" of $\mathfrak{L}_C$. If A is the lattice of pairs of nonnegative integers (r, s) such that $r \leq s$ and $s \geq 1$ with a componentwise order, and N^+ is the lattice of natural numbers ordered by divisibility, then the "skeleton" consisting of $\Omega_{r,s,n}$ is isomorphic to the direct product $A \times N^+$.

Every commutative variety lies between certain varieties from the "skeleton," namely,

$$\Omega_{V(R),L(R),D(R)} \subseteq R \subseteq \Omega_{V(R),U(R),D(R)},$$

and these inclusions are the best possible in the sense that if $\Omega_{r,s,n} \subseteq R$, then $\Omega_{r,s,n} \subseteq \Omega_{V(R),L(R),D(R)}$, and if $R \subseteq \Omega_{r,s,n}$, then $\Omega_{V(R),U(R),D(R)} \subseteq \Omega_{r,s,n}$. (Since every commutative variety is contained in a certain $\Omega_{r,s,n}$, it follows that the variety of all commutative semigroups is not a join of finitely many proper subvarieties. This result has been independently obtained by Dean and Evans [27].)

For every prime p the interval $[\Omega_{r,s,n}; \Omega_{r,s,pn}]$ does not contain any varieties except its endpoints. The same holds for the intervals $[\Omega_{r,s,n}; \Omega_{r+1,s,n}]$ for all $r \geq 0$. If $s < t$, then the intervals $[\Omega_{r,s,n}; \Omega_{r,t,n}]$ contain infinite chains.

Nelson studies certain mappings of the intervals of the "skeleton," and she studies lattices of the varieties $R_n = \{R \in \mathfrak{L}_C | D(R) = n\}$ and their interrelations for various n.

4.6. As we mentioned in 1.20, if a commutative semigroup variety has only finitely many subvarieties, then it is contained in the variety

$$V_n^k = \Pi(xy = yx,\ x_1x_2\cdots x_n = x_1x_2\cdots x_n^{k+1})$$

for some n and k. There exist descriptions of the lattices of subvarieties of V_n^k and $\overline{V}_n^k = \Pi(xy = yx,\ x_1\cdots x_n = x_1\cdots x_n y)$ (see [84]). Let N^k be the lattice of divisors of k. Then it is possible to define a congruence ρ on $\mathfrak{L}_{\overline{V}_n^k}$ such that $\mathfrak{L}_{\overline{V}_n^k}/\rho \cong N^k$ and every ρ-class is isomorphic to a sublattice of the lattice of commutative n-nilpotent semigroups. The lattice $\mathfrak{L}_{V_n^k}$ is isomorphic to a reduplication (cf. 3.1) of $\mathfrak{L}_{\overline{V}_n^k}$. In the same paper [84] all covers of the varieties V_n^k and $\overline{V}_n^k$ are found.

4.7. There exist descriptions of certain other sublattices of $\mathfrak{L}$ (see, for example, [95] and [36]).

§5. Semigroups which are free in varieties

5.1. A definition of a semigroup free in a class of semigroups is well known (see, for example, [77]). A semigroup $F \in \Pi(\Phi)$ is free in the variety $\Pi(\Phi)$ if and only if it has a generating set with all defining relations being identities which are satisfied in $\Pi(\Phi)$, i.e., they follow from Φ. It follows that two free semigroups with generating sets of the same cardinality are isomorphic. Therefore, we denote by $F_\lambda(\Phi)$ the free semigroup of a variety with a generating set of cardinality λ.

5.2. Obviously, the properties of $F_\lambda(\Phi)$ depend on those of Φ very substantially. Quite a few of the properties of $F_\lambda(\Phi)$ hold for all semigroups from $\Pi(\Phi)$. For example, any properties preserved under homomorphisms do so.

5.3. It is easy to see that $F_\lambda(\Phi)$ (for an arbitrary λ) is periodic if and only if $D(\Phi) \neq 0$, i.e. when Φ contains an unbalanced identity. In particular, $F_\lambda(\Phi)$ is idempotent if and only if $D(\Phi) = 1$ and Φ contains an unbalanced identity of the form $x = u$ [12].

5.4. The semigroup $F_\lambda(\Phi)$ ($\lambda \geq 2$) is inverse if and only if Φ contains identities of the form $x = u$ ($x \not\equiv u$), $xu_1 = yu_2$, and $v_1x = v_2y$, where u_1, u_2, v_1, and v_2 are either words or empty symbols. A semigroup $F_\lambda(\Phi)$, $\lambda \geq 2$, is a group if and only if Φ, together with the identities just mentioned, contains an anomalous identity [10].

5.5. The facts discussed in 4.3 and 4.4 can be stated differently if we use the atoms of $\mathfrak{L}$. Namely, $F_\lambda(\Phi)$ is periodic if and only if the variety $\Pi(\Phi)$ contains only finitely many atoms. It is clear that all semigroups of $\Pi(\Phi)$ are periodic in such a case. A semigroup $F_\lambda(\Phi)$, $\lambda \geq 2$, is inverse (and, therefore, all the semigroups in $\Pi(\Phi)$ are inverse) if and only if $\Pi(\Phi)$ does not contain the atoms $\Pi(xy = x)$, $\Pi(xy = y)$, and $\Pi(x_1x_2 = x_3x_4)$. If it does not contain these atoms and the atom $\Pi(x^2 = x, xy = yx)$, then all the semigroups of the variety are groups. A slightly different condition for all semigroups of a variety to be groups is contained in Evans [36].

5.6. Biryukov [12], [13] obtained quite a few results of that type. In particular, he found conditions for nilpotency of $F_\lambda(\Phi)$, $\lambda \geq 2$, for its being a semigroup with a single idempotent, or a completely simple semigroup without zero, etc.

5.7. It is interesting to describe the structure of free semigroups of varieties. We know the structure of the semigroups which are free in the varieties of all semigroups, of all idempotent semigroups, and of all commutative semigroups (see [26] and [77]).

5.8. The structure of free semigroups $F_\lambda(\Phi)$ for a finite λ is known in the variety $\Pi(\Phi)$ generated by all completely simple semigroups in which $\mathcal{H}$-classes are abelian groups [63]. Define a binary relation ρ in a free semigroup F_n of finite rank n as follows: $(u, v) \in \rho$ if the words u and v have the same first letters and the same last letters, and also if the number of occurrences of any two-letter word is the same in u and in v. It is easily seen that ρ is a congruence. The semigroups F_n/ρ are free in $\Pi(\Phi)$. This variety is finitely based, and can be determined by the following system of six balanced identities:

$$\{xyx^2 = x^2yx,\ xyxzx = xzxyx,\ xzyxy = xyxzy,$$
$$xz_1yxz_2y = xz_2yxz_1y,\ xz_1yz_2xy = xyz_2xz_1y,$$
$$xz_1yz_2xz_3y = xz_3yz_2xz_1y\}.$$

5.9. There exist results concerning the word problem in free semigroups of certain varieties. Here are some of them. This problem can be solved in an obvious way in the variety of all semigroups. It has been solved in the variety of all idempotent semigroups ([54], [90]; see also [77]). Quite recently, the word problem was solved for free semigroups of varieties determined by a single identity of the form $(x_1 \cdots x_m)^2 = x_1 \cdots x_m$, where m is any natural number [49]. On the other hand, there is an example of a finitely based variety with an unsolvable word problem for the free semigroup of countably infinite rank [103].

5.10. There exists a well-known and interesting characterization of free semigroups in the variety of all semigroups in terms of divisibility (see [26] and [77]). Mikhaĭlov [101] found an abstract characterization of free semigroups of the variety $\Pi(x_1x_2x_3x_4x_3 = x_1x_2x_4x_3^2)$; it was expressed in terms of divisibility. This variety is interesting because it consists precisely of the semigroups in which every groupoid with an infix multiplication is a semigroup.

5.11. For certain varieties the problem of lattice determinability, i.e., the determinability of their free semigroups by their subsemigroup lattices, is solved. All needed definitions can be found in [77]. Petropavlovskaya solved this problem for free semigroups in the variety of all semigroups. They are determined by their subsemigroup lattice in the class of all semigroups (see [77] and the bibliography given there). The same holds for free semigroups in the variety of all commutative semigroups, in all varieties of idempotent semigroups except finitely many such varieties (see [127]–[129]).

5.12. Every semigroup which is free in a variety of n-nilpotent semigroups ($n \geq 6,\ n \neq 7$) is lattice-determinable in the class of all semigroups. This does not hold for the remaining n [38]. It turned out that any semigroup which is free in any supracommutative variety is lattice determinable in the class of all semigroups [109]. In many of these cases the lattice isomorphisms are described as well.

5.13. A semigroup variety is called *Schreier* if every subsemigroup with more than one element in any free semigroup of the variety is free in this variety. All Schreier varieties of semigroups are exhausted by the following list:

$$\Pi(xy = x), \qquad \Pi(xy = y),$$
$$\Pi(x_1x_2 = x_3x_4), \qquad \Pi(xy = yx,\ x^py = y)$$

(p is a prime), i.e. these are precisely all the atoms but one: $\Pi(xy = yx,\ x^2 = x)$ (see [34]). The Schreier varieties are discussed in somewhat more detail by Evans [36].

5.14. As we have seen, as a rule semigroup varieties are not Schreier. Consider the following generalization of the notion of Schreier variety. A subsemigroup B of a semigroup A is called a *retract* if there exists a homomorphism of A onto B which is the identity mapping on B. A variety of semigroups is called *Hall* if any retract with more than one element of any free semigroup of the variety is free in this variety. Clearly, every Schreier variety is a Hall variety. However, the class of Hall varieties is considerably wider than the class of Schreier varieties.

Hall varieties of semigroups are precisely the following ones: Schreier varieties, supracommutative varieties, varieties consisting only of nilsemigroups, and Hall group varieties of exponent p^k for a prime p [140]. We have already mentioned that the varieties consisting of nilsemigroups form a continuous lattice (see 1.14). Supracommutative varieties form a continuous lattice as well (see 1.5). Thus, there exists a continuum of Hall varieties.

5.15. Let us recall that a semigroup A *generates* a variety V if the variety is determined by a system of identities which hold for A. It is clear that a free semigroup of a countably infinite rank always generates the variety it belongs to. For free semigroups of finite rank this is not always true. On the other hand, it is clear that if a semigroup generates a variety, then it need not be free in it. Thus we arrive at a problem: in which varieties is every semigroup generating the variety free in it? The first to raise this problem was Schein, who considered idempotent semigroups. This is the answer: every semigroup generating a variety is free in the variety precisely when the variety is an atom, except for the atom $\Pi(xy = yx,\ x^2 = x)$. These are all Schreier semigroup varieties [37].

5.16. Certain other questions related to semigroups free in varieties are discussed in Evans' survey [36]. In particular, there, and also in [33] and [35], information is given on interconnections between V being generated by finite algebras and the residual finiteness and the Hopf property of the free semigroups in V.

§6. Groupoid of varieties

6.1. Mal′cev in [83] introduced an operation of multiplication of classes of algebras. Suppose that V_1 and V_2 are subclasses of a fixed class V of algebras. The product $V_1 \cdot V_2$ is defined as the class of all algebras A in V which have congruences ρ such that all ρ-classes which are subalgebras belong to V_1 and the quotient algebras A/ρ belong to V_2.

If V is the variety of all groups, then the product of any two varieties is a variety. For semigroups this is not so, i.e. a product of two varieties need not be a variety. Therefore, it is natural to consider the following operation on the set of all semigroup varieties: the product $V_1 * V_2$ of semigroup varieties V_1 and V_2 in the variety of all semigroups is the variety generated by the class $V_1 \cdot V_2$ (this operation was mentioned in [83]).

6.2. The set $\mathfrak{L}$ of all semigroup varieties is a groupoid under the operation $*$ (this operation is not associative for semigroup varieties). Only the first steps in studying this groupoid have been made.

6.3. It has been shown in [142] how, if one knows all the identities which hold for V_1 and V_2, to obtain all identities holding in $V_1 * V_2$. As a corollary, there was found an example of two finitely based varieties V_1 and V_2 such that $V_1 * V_2$ was not finitely based. These are the variety of all commutative semigroups and the variety of all semilattices. These varieties are not just finitely based, they are hereditarily finitely based.

It was proved in the same paper that, among supracommutative varieties, there are no varieties (except the variety of all semigroups) which are closed under all (all commutative, all rectangular) bands of subsemigroups. This problem is closely connected with products of varieties: if V_1 is an arbitrary variety of semigroups and V_2 an arbitrary variety of idempotent semigroups, then $V_1 * V_2$ is precisely the variety generated by all bands of semigroups from V_1 such that the corresponding quotient semigroups belong to V_2. Certain varieties closed under bands of semigroups are known; however, there is no general description of such varieties.

6.4. The idempotents of the groupoid $\mathfrak{L}$ were found in [88]. The subgroupoid of all idempotent varieties has been described in full (by means of a multiplication table) [143]. In that paper it is mentioned that the set of all supracommutative varieties forms a subgroupoid which is a commutative semigroup with zero multiplication. Varieties Z_n of n-nilpotent semigroups form a subgroupoid which is a free commutative semigroup Z_p with free generators, where p is a prime number.

6.5. Suppose that V_1 is an arbitrary supracommutative variety and V_2 is a variety of all (all commutative, all rectangular, respectively) commutative semigroups. The chain $V_1 \subset V_1 * V_2 \subset (V_1 * V_2) * V_3 \subset \cdots$ was studied in [142]. It is infinite, and the join of its terms is the variety of all semigroups. Each interval of this chain is infinite.

6.6. The set $\mathfrak{L}$ of all semigroup varieties can be endowed with another multiplication which uses the wreath product of semigroups.

Let K be a class of semigroups. Let $S(K)$, $C(K)$, and $Q(K)$ denote the closures of K with respect to subsemigroups, direct products, and homomorphic images. Suppose that V_1 and V_2 are semigroup varieties and $K = \{A \text{ wr } B | A \in V_1,\ B \in V_2\}$, where A wr B is a wreath product of A and B. Then the product of V_1 and V_2 is defined as the class $V_1 \circ V_2 = \mathrm{QS}(K)$. This operation was first considered on the set of semigroup varieties in [73] and [74].

6.7. The multiplication we have just introduced is not associative. A corresponding example was constructed in [73]. However an analogous operation on the set of varieties of monoids is associative, i.e., the set of varieties of monoids is a semigroup with respect to this operation. Some information on the structure of this semigroup is available in [73]. The set of all supracommutative varieties of monoids is the greatest two-sided completely isolated ideal with zero multiplication in the semigroup of all varieties of monoids, and each element outside this ideal (except the trivial one) has infinite order.

Problems

When one studies and compares the results obtained while considering the lattice of semigroup varieties, one ses that a number of natural related problems have no answers. We find it appropriate to draw attention to some of them—sometimes quite narrow ones but always of indisputable interest.

The question about the cardinality of $\mathfrak{L}_\Phi$ for various systems Φ is very interesting. In particular, can $\mathfrak{L}_\Phi$ be countably infinite if Φ is an infinite irreducible system of identities? If Φ consists of a single identity, then $\mathfrak{L}_\Phi$ is "almost always" continuous (see 1.9). What can be said about the cardinality of $\mathfrak{L}_\Phi$ in the case when Φ is finite and has more than one element?

Since we have an example of a six-element semigroup without a finite base of identities, there is a problem of the minimality of such a semigroup. This problem is very close to its solution: it remains to consider noncommutative nonidempotent semigroups having no zero or identity (see 1.15). Probably, it is time for the problem of finding some interesting classes of semigroups in which every finite semigroup (but not every semigroup) has a finite base of identities. An inspiring example is provided, as we know, by the class of groups. Also, it is noteworthy that the six-element semigroup just mentioned is inverse.

The problem of Evans on describing "small" varieties of semigroups is not yet solved completely; however, it is partly solved (see 1.19 and 1.21). Thus, a few narrower problems arise. Must a variety be "small" if it satisfies a permutational identity and an unbalanced identity of the form $x_1 \cdots x_n = v$? If an unbalanced identity $x_1 \cdots x_n = v$ does not hold in a variety V, the variety is not "small." Moreover, the lattice $\mathfrak{L}_V$ has an infinite ascending chain (cf. 1.18). What is the role of a permutational identity? There are examples of "small" varieties in which no permutational identity holds. However, every proper subvariety of such varieties satisfies a permutational identity. Therefore, the following problem emerges: is there a variety such that its lattice of subvarieties is infinite and satisfies both the maximality and minimality conditions?

If a variety V is determined by a system of balanced identities, then every variety in $\mathfrak{L}_V$ has a cover (cf. 2.1). This is an answer to the basic question on covers in the lattice of semigroup varieties raised by Evans. However, quite a few questions about covers in this lattice remain unsolved. In 2.3 we gave an example of a variety V which, for a variety U which properly contains V, has no cover in $\mathfrak{L}_U$. We mentioned there that V had no irreducible base of identities. Is this fact accidental? For which varieties can such "overvarieties" exist (clearly, not for all varieties)?

Suppose that V is a variety and Ω is the set of all atoms contained in it. Let $\overline{\mathfrak{L}}_\Omega$ be the lattice of all such varieties which do not contain any atoms not from Ω (see 3.11). Finding the covers of V in $\overline{\mathfrak{L}}_\Omega$ is a very interesting problem. Indeed, among the varieties which do not belong to this lattice, only the varieties of the form $V \vee \Gamma$, where Γ is an atom not from Ω, can cover V. For which varieties do all their covers belong or, conversely, not belong to $\overline{\mathfrak{L}}_\Omega$? Which varieties have a finite, or countably infinite, set of covers? Which have continuum many covers? What is the cardinality of the set of varieties which have finitely many covers? Countably many covers? We know that there exist a continuum of varieties each of which has a continuum of covers (see 2.5).

Biryukov [16] described minimal noncommutative varieties consisting not only of groups, i.e., noncommutative varieties all proper subvarieties of which are commutative. The following problem seems very interesting: which are the minimal varieties satisfying no permutational identities? A solution to this problem may be very useful in finding "small" varieties.

REFERENCES

1. S. I. Adyan, *Identities in special semigroups*, Dokl. Akad. Nauk SSSR **143** (1962), 499–502 = Soviet Math. Dokl. **3** (1962), 401–404. MR **26** #3806.

2. A. Ya. Aĭzenshtat, *Coverings of nilpotent varieties of semigroups*, Modern Algebra. Groupoids and Their Homomorphisms (E. S. Lyapin, editor), Leningrad. Gos. Ped. Inst., Leningrad, 1980, pp. 3–11. (Russian) MR **81m:** 20075.

3. ——, *On coverings in lattices of varieties of semigroups*, Modern Analysis and Geometry [= Leningrad. Gos. Ped. Inst. Uchen. Zap. **541**; I. Ya. Bakel′man et al., editors], Leningrad. Gos. Ped. Inst., Leningrad, 1972, pp. 3–11. (Russian) R. Zh. Mat. **1973**, 5A 164.

4. ——, *Certain sublattices of the lattice of varieties of semigroups*, Modern Algebra, Vyp. 1 (E. S. Lyapin, editor), Leningrad. Gos. Ped. Inst., Leningrad, 1974, pp. 3–15. (Russian) MR **54** #432.

5. ——, *Permutation identities*, Modern Algebra, Vyp. 3 (E. S. Lyapin, editor), Leningrad. Gos. Ped. Inst., Leningrad, 1975, pp. 3–12. (Russian) MR **56** #3155.

6. ——, *The lattice of varieties of semigroups that contain no atoms other than* $\Pi(x_1x_2 = y_1y_2)$ *and* $\Pi(xy = x)$, Modern Algebra, Vyp. 6 (E. S. Lyapin and M. M. Lesokhin, editors), Leningrad. Gos. Ped. Inst., Leningrad, 1977, pp. 3–15. (Russian) MR **58** #28225.

7. ——, *On varieties of semigroups having a finite number of subvarieties*, Algebraic Theory of Semigroups (Proc. Sixth Algebraic Conf., Szeged, 1976; G. Pollák, editor), Colloq. Math. Soc. János Bolyai, vol. 20, North-Holland, 1979, pp. 33–41, MR **80k:** 20057.

8. A. K. Austin, *A closed set of laws which is not generated by a finite set of laws*, Quart. J. Math. Oxford Ser. (2) **17** (1966), 11–13. MR **34** #4341.

9. R. A. Baĭramov, A. A. Makhmudov, and R. B. Feĭzullaev, *Some finiteness conditions and operators in the theory of varieties*, Izv. Akad. Nauk Azerbaĭdzhan. SSR Ser. Fiz.-Tekhn. Mat. Nauk **1977**, no. 5, 93–98. (Russian) MR **58** #21896.

10. John T. Baldwin and Joel Berman, *Varieties and finite closure conditions*, Colloq. Math. **35** (1976), 15–20. MR **53** #5427.

11. A. P. Biryukov, *On infinite assemblies of identities in semigroups*, Algebra i Logika **4** (1965), no. 2, 31–32. (Russian) MR **35** #2980.

12. ——, *Semigroups given by identities*, Novosibirsk. Gos. Ped. Inst. Uchen. Zap. **18** (1963), 139–169. (Russian) Cf. also his short note in Dokl. Akad. Nauk SSSR **149** (1963), 230–232 = Soviet Math. Dokl. **4** (1963), 333–336. MR **26** #6281.

13. ——, *On some identities in semigroups*, Novosibirsk. Gos. Ped. Inst. Uchen. Zap. **18** (1963), 170–184. (Russian) R. Zh. Mat. **1964**, 5A 193.

14. A. P. Birjukov, *Semigroups given by identities with distinguished elements*, Mat. Sb. **67(109)** (1965), 461–473; English transl., Amer. Math. Soc. Transl. (2) **82** (1969), 209–223. MR **33** #2740.

15. ——, *Varieties of idempotent semigroups*, Algebra i Logika **9** (1970), 255–273 = Algebra and Logic **9** (1970), 153–164. MR **45** #6949.

16. ——, *Minimal noncommutative varieties of semigroups*, Sibirsk. Mat. Zh. **17** (1966), 677–681 = Siberian Math. J. **17** (1976), 520–522. MR **53** #13441.

17. B. K. Boguta, *Coverings of varieties defined by one identity*, Modern Algebra (E. S. Lyapin, editor), Leningrad. Gos. Ped. Inst., Leningrad, 1978, pp. 24–28. (Russian) Zbl **498** #20040.

18. B. K. Boguta and A. M. Nikolaev, *Coverings in the lattice of supracommutative varieties of semigroups*, Fourteenth All-Union Algebra Conf., Abstracts of Reports, Vol. I, Novosibirsk. Gos. Univ., Novosibirsk, 1977. (Russian)

19. A. D. Bol′bot, *Finiteness of the base of identities of four-element semigroups*, Eleventh All-Union Algebra Colloq., Abstracts of Reports, "Shtiintsa", Kishinev, 1971. (Russian)

20. Thomas C. Brown, *On the finiteness of semigroups in which* $x^r = x$, Proc. Cambridge Philos. Soc. **60** (1964), 1028–1029. MR **29** #3562.

21. Stanley Burris and Evelyn Nelson, *Embedding the dual of* Π_m *in the lattice of equational classes of commutative semigroups*, Proc. Amer. Math. Soc. **30** (1971), 37–39. MR **44** #2857.

22. S. Y. Chen and Simon C. Hsieh, *A note on rectangular semigroups*, Tamkang J. Math. **4** (1973), 27–28. MR **48** #11364.

23. J. L. Chrislock, *On medial semigroups*, J. Algebra **12** (1969), 1–9. MR **38** #5966.

24. ——, *A certain class of identities on semigroups*, Proc. Amer. Math. Soc. **21** (1969), 189–190. MR **38** #5954.

25. David M. Clark, *Disassociative groupoids are not finitely based*, J. Austral. Math. Soc. **11** (1970), 113–114. MR **41** #5531.

26. A. H. Clifford and G. B. Preston, *The algebraic theory of semigroups*, Vols. I, II, Amer. Math. Soc., Providence, R.I., 1961, 1967. MR **24** #A2627; **36** #1558.

27. Richard A. Dean and Trevor Evans, *A remark on varieties of lattices and semigroups*, Proc. Amer. Math. Soc. **21** (1969), 394–396. MR **39** #2849.

28. G. G. Dyadchenko, *Identities on monogenic inverse semigroups*, Algebra and Number Theory, vyp. 2 (Kh. Ya. Unachev, editor), Kabardino-Balkarsk. Gos. Univ., Nalchik, 1977, pp. 57–77. (Russian) MR **57** #16436.

29. Charles C. Edmunds, *On certain finitely based varieties of semigroups*, Semigroup Forum **15** (1977/78), 21–39. MR **57** #3286.

30. Samuel Eilenberg and M. P. Schützenberger, *On pseudovarieties*, Advances in Math. **19** (1976), 413–418. MR **53** #5431.

31. Trevor Evans, *A condition for a cancellation semigroup to be a group*, Amer. Math. Monthly **73** (1966), 1104–1106. MR **36** #2722.

32. ——, *The number of semigroup varieties*, Quart J. Math. Oxford Ser. (2) **19** (1968), 335–336. MR **38** #257.

33. ——, *Some connections between residual finiteness, finite embeddability and the word problem*, J. London Math. Soc. (2) **1** (1969), 399–403. MR **40** #2589.

34. ——, *Schreier varieties of semigroups*, Math. Z. **112** (1969), 296–299. MR **40** #2776.

35. ——, *Finitely presented loops, lattices, etc. are Hopfian*, J. London Math. Soc. **44** (1969), 551–552. MR **38** #5975.

36. ——, *The lattice of semigroup varieties*, Semigroup Forum **2** (1971), 1–43. MR **44** #1753.

37. ——, *When is a functionally free algebra free? A remark on a problem of B. M. Schein*, Semigroup Forum **4** (1972), 177–184. MR **45** #417.

38. A. E. Evseev, *Lattice isomorphisms of free nilpotent semigroups*, Leningrad. Gos. Ped. Inst. Uchen. Zap. **387** (1968), 112–125. (Russian) MR **37** #4184.

39. Charles F. Fennemore, *All varieties of bands.* I, II, Math. Nachr. **48** (1971), 237–252, 253–262. MR **45** #3605.

40. ——, *Characterization of bands satisfying no non-trivial identity*, Semigroup Forum **2** (1971), 371–375. MR **44** #2858.

41. ——, *A subdirectly irreducible band accepting* $a = a$, Semigroup Forum **9** (1974/75), 271–274. MR **51** #10494.

42. J. Froemke and R. Quackenbush, *The spectrum of an equational class of groupoids*, Pacific J. Math. **58** (1975), 381–386. MR **52** #5522.

43. László Fuchs, *A remark on lattice-ordered semigroups*, Semigroup Forum **7** (1974), 372–374. MR **52** #5510.

44. E. Ya. Gabovich and M. S. Trepetin, *Varieties of nilpotent semigroups*, Tartu Riikl. Ül. Toimetised No. 277 (1971), 16–19. (Russian) MR **52** #602.

45. J. A. Gerhard, *The lattice of equational classes of idempotent semigroups*, J. Algebra **15** (1970), 195–224. MR **41** #8552.

46. ——, *Subdirectly irreducible idempotent semigroups*, Pacific J. Math. **39** (1971), 669–676. MR **47** #1980.

47. ——, *Some subdirectly irreducible idempotent semigroups*, Semigroup Forum **5** (1972/73), 362–369. MR **50** #517.

48. ——, *Injectives in equational classes of idempotent semigroups*, Semigroup Forum **9** (1974/75), 36–53. MR **52** #535.

49. ——, *Semigroups with an idempotent power*. I: *Word problems*, Semigroup Forum **14** (1977), 137–141. MR **58** #5990.

50. ——, *Semigroups with an idempotent power*. II: *The lattice of equational subclasses of* $[(xy)^2 = xy]$, Semigroup Forum **14** (1977), 375–388. MR **58** #5991.

51. J. A. Gerhard and Ahmad Shafaat, *Semivarieties of idempotent semigroups*, Proc. London Math. Soc. (3) **22** (1971), 667–680. MR **45** #2048.

52. Max Dieter Gerhardts, *Termgleichungen in glatt-idempotenten Halbgruppen*, Arch. Math. (Basel) **19** (1968), 289–290. MR **37** #4180.

53. Zensiro Goseki, *On semigroups in which* $xyx = xy$, Sci. Rep. Fac. Ed. Gunma Univ. **22** (1973), 5–10. R. Zh. Mat. **1975**, 3A215.

54. J. A. Green and D. Rees, *On semi-groups in which* $x^r = x$, Proc. Cambridge Philos. Soc. **48** (1952), 35–40. MR **13**, 720.

55. E. I. Grindlinger, *On the question of identities in semigroups*, Izv. Vyssh. Uchebn. Zaved. Mat. **1967**, no. 5(60), 6–12. (Russian) MR **35** #6769.

56. T. E. Hall, *On bands, free bands and identities*, Semigroup Forum **2** (1971), 83–85. MR **44** #1749.

57. T. J. Head, *The varieties of commutative monoids*, Neiuw Arch. Wisk. (3) **16** (1968), 203–206. MR **38** #5977.

58. V. I. Igoshin, *Remarks on characterizable classes of algebras with examples from the theory of semigroups*, Theory of Semigroups and Its Applications, vyp. 3 (V. V. Vagner, editor), Izdat. Saratov. Univ., Saratov, 1974, pp. 30–39. (Russian) MR **56** #15535.

59. John R. Isbell, *Two examples in varieties of monoids*, Proc. Cambridge Philos. Soc. **68** (1970), 265–266. MR **41** #8560.

60. Jaroslav Ježek, *Primitive classes of algebras with unary and nullary operations*, Colloq. Math. **20** (1969), 159–179. MR **40** #82.

61. ——, *Intervals in the lattice of varieties*, Algebra Universalis **6** (1976), 147–158. MR **54** #7354.

62. Jan Kalicki and Dana Scott, *Equational completeness of abstract algebras*, Nederl. Akad. Wetensch Proc. Ser. A **58** = Indag. Math. **17** (1955), 650–659. MR **17**, 571.

63. Ki Hang Kim and F. Roush, *The semigroup of adjacency patterns of words*, Algebraic Theory of Semigroups (Proc. Sixth Algebraic Conf., Szeged, 1976; G. Pollák, editor), Colloq. Math. Soc. János Bolyai, vol. 20, North-Holland, 1979, pp. 281–297. MR **80h**: 20083.

64. Naoki Kimura, *Note on idempotent semigroups*. IV, Proc. Japan Acad. **34** (1958), 121–123. MR **20** #4604.

65. ——, *The structure of idempotent semigroups*. I, Pacific J. Math. **8** (1958), 257–275.

66. E. I. Kleĭman, *Free inverse semigroups*, Ural. Gos. Univ. Mat. Zap. **8** (1972), no. 3, 49–72. (Russian) MR **48** #4160.

67. ——, *The lattice of varieties of inverse semigroups*, Izv. Vyssh. Uchebn. Zaved. Mat. **1976**, no. 7(170), 106–109 = Soviet Math. (Iz. VUZ) **20** (1976), no. 7, 86–88. MR **58** #10662.

68. ——, *On basis of identities of Brandt semigroups*, Semigroup Forum **13** (1976/77), 209–218. MR **56** #509.

69. Miroslav I. Klun, *A characterization of varieties of inverse semigroups*, Semigroup Forum **10** (1975/76), 1–7. MR **52** #5834.

70. T. L. Kolesnikova, *Semigroups with a single defining relation in the class of strongly nilpotent semigroups of degree two*, Modern Algebra, Vyp. 2 (E. S. Lyapin, editor), Leningrad. Gos. Ped. Inst., Leningrad, 1974, pp. 16–23. (Russian) MR **53** #10954.

71. ——, *Identities in restrictive semigroups with a single defining relation*, Modern Algebra, Vyp. 2 (E. S. Lyapin, editor), Leningrad. Gos. Ped. Inst., Leningrad, 1974, pp. 24–34. (Russian) MR **57** #495.

72. Yu. G. Koshelev, *Varieties closed under wreath products*, Twenty-Eighth Hercen Lectures: Mathematics, Summaries of Papers, Leningrad. Gos. Ped. Inst., Leningrad, 1975, pp. 7–11. (Russian) R. Zh. Mat. **1975**, 10A186.

73. ——, *An associative operation on the set of all varieties of monoids*, Modern Algebra, Vyp. 4 (E. S. Lyapin, editor), Leningrad. Gos. Ped. Inst., Leningrad, 1976, pp. 107–117. (Russian) MR **57** #504.

74. ——, *Varieties preserved under wreath products*, Semigroup Forum **12** (1976), 95–107. MR **53** #3175.

75. V. M. Krivenko, *Hamiltonian varieties of semigroups*, Thirtieth Hercen Lectures: Mathematics, Summaries of Papers, Leningrad. Gos. Ped. Inst., Leningrad, 1977, pp. 9–11. (Russian) R. Zh. Mat. **1977**, 9A202.

76. S. I. Kublanovskiĭ, *Finitely separable prevarieties of semigroups*, Thirty-First Hercen Lectures: Mathematics, Summaries of Papers, Leningrad. Gos. Ped. Inst., Leningrad, 1978, pp. 25–28. (Russian) R. Zh. Mat. **1978**, 10A120.

77. E. S. Lyapin, *Semigroups*, Fizmatgiz, Moscow, 1960; English transl., Amer. Math. Soc., Providence, R.I., 1963. MR **22** #11054; **29** #4817.

78. ——, *Semigroups*, 3rd rev. ed. of English transl., Amer. Math. Soc., Providence, R.I., 1974. MR **50** #4789.

79. ——, *An infinite irreducible assembly of semigroup identities*, Mat. Zametki **7** (1970), 545–549 = Math. Notes **7** (1970), 330–332. MR **42** #396.

80. ——, *On imbedding semigroup identities in infinite irreducible assemblies*, Mat. Zametki **12** (1972), 95–104 = Math. Notes **12** (1972), 489–494. MR **47** #6896.

81. F. F. Lysenko, *An infinite irreducible assembly of anomal identities*, Thirty-First Hercen Lectures: Mathematics, Summaries of Papers, Leningrad. Gos. Ped. Inst., Leningrad, 1978, pp. 29–32. (Russian) R. Zh. Mat. **1978**, 12A189.

82. A. I. Mal'tsev, *Algebraic systems*, "Nauka", Moscow, 1970; English transl., Springer-Verlag, 1973.

83. ——, *Multiplication of classes of algebraic systems*, Sibirsk. Mat. Zh. **8** (1967), 346–365 = Siberian Math. J. **8** (1967), 254–267. MR **35** #4140.

84. S. A. Malyshev, *The lattice of subvarieties of the variety* $\Pi(x_1\cdots x_n = x_1\cdots x_n^{k+1}, xy = yx)$, Semigroup Varieties and Semigroups of Endomorphisms (E. S. Lyapin, editor), Leningrad. Gos. Ped. Inst., Leningrad, 1979, pp. 113–122. (Russian) MR **81i:** 08011.

85. S. G. Mamikonyan, *Varieties of residually finite semigroups*, Mat. Sb. **88(130)** (1972), 353–359 = Math. USSR Sb. **17** (1972), 349–355. MR **47** #379.

86. ——, *Varieties of residually finite semigroups*. II, Modern Algebra, Vyp. 6 (E. S. Lyapin and M. M. Lesokhin, editors), Leningrad Gos. Ped. Inst., Leningrad, 1977, pp. 136–140. (Russian) MR **58** #28228.

87. L. M. Martynov, *On attainable classes of groups and semigroups*, Mat. Sb. **90(122)** (1973), 235–245 = Math. USSR Sb. **19** (1973), 231–241. MR **51** #13111.

88. ——, *On varieties of semigroups closed under ideal extensions*, Nauchn. Trudov. Novosibirsk. Gos. Ped. Inst. Vyp. 94 (1974), 72–73. (Russian) R. Zh. Mat. **1975**, 1A223.

89. G. I. Mashevitskiĭ, *On identities of completely simple semigroups over Abelian groups*, Thirty-First Hercen Lectures: Mathematics, Summaries of Papers, Leningrad. Gos. Ped. Inst., Leningrad, 1978, pp. 33–35. (Russian) R. Zh. Mat. **1978**, 12A190.

90. David McLean, *Idempotent semigroups*, Amer. Math. Monthly **61** (1954), 110–113. MR **15**, 681.

91. D. G. Mead and Takayuki Tamura, *Semigroups satisfying* $xy^m = yx^m = (xy^m)^n$, Proc. Japan Acad. **44** (1968), 779–781. MR **38** #5960.

92. I. I. Mel'nik, *Varieties of* Ω*-algebras*, Studies in Algebra, Vyp. 1 (V. V. Vagner, editor), Izdat. Saratov. Univ., Saratov, 1969, pp. 32–40. (Russian) MR **40** #7179.

93. ——, *Varieties and lattices of varieties of semigroups*, Studies in Algebra, Vyp. 2 (V. V. Vagner, editor), Izdat. Saratov. Univ., Saratov, 1970, pp. 47–57. (Russian) MR **54** #433.

94. ——, *A family of varieties of semigroups*, Izv. Vyssh. Uchebn. Zaved. Mat. **1971**, no. 12 (115), 103–108. (Russian) MR **45** #6963.

95. ——, *A description of certain lattices of varieties of semigroups*, Izv. Vyssh. Uchebn. Zaved. Mat. **1972**, no. 7 (122), 65–74. (Russian) MR **51** #10173.

96. ——, *Nilpotent translations of varieties*, Mat. Zametki **14** (1973), 703–712 = Math. Notes **14** (1973), 962–966. MR **51** #3028.

97. ——, *Majorant translations of varieties*, Studies in Algebra, Vyp. 4 (V. V. Vagner, editor), Izdat. Saratov. Univ., Saratov, 1974, pp. 70–78. (Russian) MR **58** #5998.

98. ——, *Regular closures of semigroup varieties endowed with a commuting idempotent*, Studies in Algebra, Vyp. 2 (V. V. Vagner, editor), Izdat. Saratov. Univ., Saratov, 1970, pp. 58–66. (Russian) MR **52** #13581.

99. ——, *On multiplication of varieties*, Theory of Semigroups and Its Applications, Vyp. 3 (V. V. Vagner, editor), Izdat. Saratov. Univ., Saratov, 1974, pp. 59–63. (Russian) Zbl. **353** #08002.

100. ——, *Majorant translations and lattices of varieties*, Ordered Sets and Lattices, Vyp. 4 (V. N. Saliĭ, editor), Izdat. Saratov. Univ., Saratov, 1977, pp. 85–98. (Russian) MR **58** #21899.

101. A. B. Mikhaĭlov, *An abstract characterization of semigroups that are free in a certain variety*, Modern Algebra, Vyp. 1 (E. S. Lyapin, editor), Leningrad. Gos. Ped. Inst., Leningrad, 1974, pp. 146–156. (Russian) MR **53** #10956.

102. ——, *Semigroups with the identity* $\xi_1\xi_2\xi_3 \cong \xi_1\xi_4\xi_3$, Twenty-Eighth Hercen Lectures: Mathematics, Summaries of Papers, Leningrad. Gos. Ped. Inst., Leningrad, 1975, pp. 11–14. (Russian) R. Zh. Mat. **1975**, 10A181.

103. V. L. Murskiĭ, *Some examples of varieties of semigroups*, Mat. Zametki **3** (1968), 663–670 = Math. Notes **3** (1968), 423–427. MR **38** #256.

104. ——, *Unrecognizable properties of finite systems of identities*, Dokl. Akad. Nauk SSSR **196** (1971), 520–522 = Soviet Math. Dokl. **12** (1971), 183–186. MR **44** #4084.

105. Evelyn Nelson, *The lattice of equational classes of semigroups with zero*, Canad. Math. Bull. **14** (1971), 531–534. MR **47** #1983.

106. ——, *The lattice of equational classes of commutative semigroups*, Canad. J. Math. **23** (1971), 875–895. MR **44** #5395.

107. A. M. Nikolaev, *On the imbedding of balanced identities in irreducible sets of identities*, Modern Algebra (E. S. Lyapin, editor), Leningrad. Gos. Ped. Inst., Leningrad, 1978, pp. 143–145. (Russian) Zbl. **463** #20041.

108. Thomas Nordahl, *Semigroups satisfying* $(xy)^m = x^m y^m$, Semigroup Forum **8** (1974), 332–346. MR **51** #10491.

109. A. Ya. Ovsyannikov, *Lattice isomorphisms with respect to free semigroups*, Mat. Zametki **21** (1977), 459–471 = Math. Notes **21** (1977), 257–264. MR **57** #12729.

110. Peter Perkins, *Bases for equational theories of semigroups*, J. Algebra **11** (1968), 298–314. MR **38** #2232.

111. Mario Petrich, *All subvarieties of a certain variety of semigroups*, Semigroup Forum **7** (1974), 104–152. MR **52** #3395.

112. ——, *Varieties of orthodox bands of groups*, Pacific J. Math. **58** (1975), 209–217. MR **52** #3405.

113. György Pollák, *On free subsemigroups of a free semigroup*, Studia Sci. Math. Hungar. **7** (1972), 317–319. MR **48** #8662.

114. ——, *On the consequences of permutation identities*, Acta Sci. Math. (Szeged) **34** (1973), 323–333. MR **48** #448.

115. ——, *On hereditarily finitely based varieties of semigroups*, Acta Sci. Math. (Szeged) **37** (1975), 339–348. MR **52** #10911.

116. ——, *A class of hereditarily finitely based varieties of semigroups*, Algebraic Theory of Semigroups (Proc. Sixth Algebraic Conf., Szeged, 1976; G. Pollák, editor), Colloq. Math. Soc. János Bolyai, vol. 20, North-Holland, 1979, pp. 433–445. MR **80j:** 08008a.

117. ——, *On identities which define hereditarily finitely based varieties of semigroups*, Algebraic Theory of Semigroups (Proc. Sixth Algebraic Conf., Szeged, 1976; G. Pollák, editor), Colloq. Math. Soc. János Bolyai, vol. 20, North-Holland, 1979, pp. 447–452. MR **80j:** 08008b.

118. Mohan S. Putcha and Julian Weissglass, *Semigroups satisfying variable identities*. I, Semigroup Forum **3** (1971/72), 64–67. MR **44** #6881.

119. ——, *Semigroups satisfying variable identities*. II, Trans. Amer. Math. Soc. **168** (1972), 113–119. MR **45** #6954.

120. Mohan S. Putcha and Adil Yaqub, *Semigroups satisfying permutation identities*, Semigroup Forum **3** (1971/72), 68–73. MR **45** #2050.

121. V. N. Saliĭ, *Equationally normal varieties of semigroups*, Izv. Vyssh. Uchebn. Zaved. Mat. **1969**, no. 5 (84), 61–68. (Russian) MR **41** #8555.

122. B. M. Schein, *An example of two non-isomorphic finite subdirectly irreducible bands generating the same variety of bands*, Semigroup Forum **4** (1972), 365–366. MR **46** #1945.

123. M. P. Schützenberger, *Sur les monoïdes finis n'ayant que des sous-groupes triviaux*, Séminaire P. Dubreil, M.-L. Dubreil-Jacotin, L. Lesieur et C. Pisot, 18e Année: 1964/65, Algèbre et Théorie des Nombres, Fasc. 1, Exposé 10, Secrétariat Math., Paris, 1967. MR **36** #1258.

124. Robert Schwabauer, *A note on commutative semigroups*, Proc. Amer. Math. Soc. **20** (1969), 503–504. MR **38** #2233.

125. ———, *Commutative semigroup laws*, Proc. Amer. Math. Soc. **22** (1969), 591–595. MR **39** #5734.

126. L. N. Shevrin, *Theory of periodic semigroups*, Izv. Vyssh. Uchebn. Zaved. Mat. **1974**, no. 5 (144), 205–215 = Soviet Math. (Iz. VUZ) **18** (1974), no. 5, 172–181. MR **50** #13336.

127. ———, *Lattice isomorphisms of commutative aperiodic semigroups with cancellation*, Mat. Sb. **63(105)** (1964), 43–58; English transl., Amer. Math. Soc. Transl. (2) **58** (1960), 77–92. MR **29** #1272.

128. ———, *Lattice properties of semigroups of idempotents*. I. Sibirsk. Mat. Zh. **6** (1965), 459–474. (Russian) MR **31** #1315.

129. ———, *Lattice properties of semigroups of idempotents*. II, Sibirsk. Mat. Zh. **7** (1966), 437–454 = Siberian Math. J. **7** (1966), 357–369. MR **33** #5772.

130. L. N. Shevrin and L. M. Martynov, *Attainable classes of algebras*, Sibirsk. Mat. Zh. **12** (1971), 1363–1381 = Siberian Math. J. **12** (1971), 986–998. MR **45** #153.

131. L. M. Shneerson, *Identities in a semigroup with a single defining relation*. I, Ivanov. Gos. Ped. Inst. Uchen. Zap. **106** (1972), 139–156. (Russian) R. Zh. Mat. **1973**, 3A187.

132. ———, *Identities in a semigroup with a single defining relation*. II, Ivanov. Gos. Ped. Inst. Uchen. Zap. **117** (1972), 112–124. (Russian) R. Zh. Mat. **1973**, 8A134.

133. ———, *Identities in finitely presented semigroups*, Ivanov. Gos. Ped. Inst. Uchen. Zap. **123** (1973), 163–201. (Russian) MR **57** #491.

134. ———, *Free subgroups of finitely presented semigroups*, Sibirsk. Mat. Zh. **15** (1974), 450–454 = Siberian Math. J. **15** (1974), 325–328. MR **48** #11362.

135. È. G. Shutov, *Semigroups with ideal subsemigroups*, Modern Algebra, Vyp. 3 (E. S. Lyapin, editor), Leningrad. Gos. Ped. Inst., Leningrad, 1975, pp. 134–158. (Russian) MR **55** #12843.

136. E. P. Simel′gor, *Identities in a finite symmetric semigroup*, Modern Algebra, Vyp. 1 (E. S. Lyapin, editor), Leningrad. Gos. Ped. Inst., Leningrad, 1974, pp. 174–188. (Russian) MR **55** #10583.

137. ———, *Subsemigroups of a symmetric semigroup that satisfy certain identities*, Modern Algebra, Vyp. 2 (E. S. Lyapin, editor), Leningrad. Gos. Ped. Inst., Leningrad, 1974, pp. 154–168. (Russian) MR **53** #3173.

138. ———, *Identities of a certain semigroup of endomorphisms of an ordered set*, Modern Algebra, Vyp. 5 (E. S. Lyapin and M. M. Lesokhin, editors), Leningrad. Gos. Ped. Inst., Leningrad, 1976, pp. 107–113. (Russian) MR **57** #6241.

139. ———, *Identities of four-element semigroups*, Modern Algebra (E. S. Lyapin, editor), Leningrad. Gos. Ped. Inst., Leningrad, 1978, pp. 146–152. (Russian) Zbl. **474** #20031.

140. L. A. Skornyakov, *Hall varieties of semigroups*, Mat. Zametki **11** (1972), 463–470 = Math. Notes **11** (1972), 281–285. MR **45** #8757.

141. H. Subramanian and T. R. Sundararaman, *Pre-complete varieties of semigroups*, Semigroup Forum **8** (1974), 278–281.

142. E. V. Sukhanov, *Varieties and bands of semigroups*, Sibirsk. Mat. Zh. **18** (1977), 419–428 = Siberian Math. J. **18** (1977), 300–307. MR **56** #15809.

143. ———, *The groupoid of varieties of idempotent semigroups*, Semigroup Forum **14** (1977), 143–159. MR **56** #12156.

144. T. L. Sundukova, *Semigroups with a single defining relation possessing a nontrivial identity*, Twenty-Sixth Hercen Lectures: Mathematics, Summaries of Papers, Leningrad. Gos. Ped. Inst., Leningrad, 1973, pp. 29–35. (Russian) R. Zh. Mat. **1973**, 10A121.

145. Takayuki Tamura, *Report on attainability of systems of identities*, Bull. Amer. Math. Soc. **71** (1965), 555–558. MR **31** #1316.

146. ——, *Attainability of systems of identities on semigroups*, J. Algebra **3** (1966), 261–276. MR **32** #7668.

147. ——, *Semigroups satisfying identity $xy = f(x, y)$*, Pacific J. Math. **31** (1969), 513–521. MR **41** #5528.

148. ——, *Complementary semigroups and exponent semigroups of order bounded groups*, Math. Nachr. **59** (1974), 17–34. MR **50** #518.

149. ——, *Note on attainability of identities on bands*, J. Algebra **28** (1974), 1–9. MR **48** #11363.

150. Takayuki Tamura and John Shafer, *On exponential semigroups*. I, Proc. Japan Acad. **48** (1972), 77–80. MR **47** #5156.

151. A. N. Trakhtman, *Covering elements in the lattice of varieties of algebras*, Mat. Zametki **15** (1974), 307–312 = Math. Notes **15** (1974), 174–177. MR **50** #4446.

152. ——, *Semigroups whose subsemigroups all belong to a certain variety*, Ural. Gos. Univ. Mat. Zap. **9** (1974), no. 1, 110–126. (Russian) MR **52** #10933.

153. ——, *Varieties of semigroups without an irreducible base of identities*, Mat. Zametki **21** (1977), 865–872 = Math. Notes **21** (1977), 487–490. MR **58** #21901.

154*. ——, *The finite basis question for semigroups of order less than six*, Semigroup Forum **27** (1983), 387–389. MR **85d:** 20070.

155. M. S. Trepetin, *A description of the semigroups that are relatively free in varieties of nilpotent semigroups*, Ukrain. Mat. Zh. **22** (1970), 697–699 = Ukrainian Math. J. **22** (1970), 603–604. MR **43** #3370.

156. Peter G. Trotter, *Exponential Δ-semigroups*, Semigroup Forum **12** (1976), 313–331. MR **54** #439.

157. E. J. Tully, Jr., *The equivalence, for varieties of semigroups, of two properties concerning congruence relations*, Bull. Amer. Math. Soc. **70** (1964), 399–400. MR **28** #4048.

158. N. S. Ulitina, *Varieties of semigroups generated by semigroups of all directed transformations of ordered sets*, Twenty-Sixth Hercen Lectures: Mathematics, Summaries of Papers, Leningrad. Gos. Ped. Inst., Leningrad, 1973, pp. 35–41. (Russian) R. Zh. Mat. **1973**, 10A123.

159. Miyuke Yamada, *Inversive semigroups*. III, Proc. Japan Acad. **41** (1965), 221–224. MR **32** #4204.

160. ——, *Note on the structure of regular semigroups*, Proc. Japan Acad. **42** (1966), 136–140. MR **33** #5776.

161. ——, *Regular semigroups whose idempotents satisfy permutation identities*, Pacific J. Math. **21** (1967), 371–392. MR **37** #2887.

162. Miyuke Yamada and Naoki Kimura, *Note on idempotent semigroups*. II, Proc. Japan Acad. **34** (1958), 110–112. MR **20** #4603.

163. A. P. Zamyatin, *Prevarieties of semigroups whose elementary theory is decidable*, Algebra i Logika **12** (1973), 417–432 = Algebra and Logic **12** (1973), 233–241. MR **48** #2631.

Translated by B. M. SCHEIN

*Added by translator.

Amer. Math. Soc. Transl.
(2) Vol. **134**, 1987

0-Cohomology of Completely 0-Simple Semigroups

UDC 512.664.4

B. V. NOVIKOV

1. The results of this paper are a generalization of a description of the cohomology groups of a completely simple semigroup obtained in [1]. The definition of the 0-cohomology of semigroups and related constructions are given in [2].

The following notation is used in the sequel: $S = S^0(G; I, \Lambda; P)$ is a completely 0-simple semigroup [3] with the basic group G and sandwich matrix $P = (p_{\lambda i})$ of size $\Lambda \times I$, e is the identity of G, A is a 0-module over S, and $\overline{S}$ is a semigroup without zero which corresponds to S [2]. We assume that $I \cap \Lambda = \{1\}$, $p_{11} = e$, and the group G is identified with G_{11}.

Since the semigroup S is categorial at 0, $H_0^n(S, A) \cong H^n(\overline{S}, A)$ by Theorem 3 of [2]. Thus our problem is reduced to that of describing the group $H^n(\overline{S}, A)$. As a preliminary we prove a few lemmas.

2. LEMMA. *Let T be an arbitrary semigroup and let G be one of its subgroups with the identity element e. Then $\mathbf{Z}Te$ is a projective right G-module.*

PROOF. Introduce a relation π on Te assuming that $se\pi te$ if and only if $sg = te$ for some $g \in G$. Since G is a group, π is an equivalence. Let $\{u_i\}$ be a set of representatives of the equivalence classes. Then $\mathbf{Z}Te = \sum_i \mathbf{Z}(u_iG)e$. The right G-modules $\mathbf{Z}(u_iG)$ and $\mathbf{Z}G$ are obviously isomorphic. Since the G-module $\mathbf{Z}G$ is projective, so is the G-module $\mathbf{Z}Te$.

3. LEMMA. *Let T be an arbitrary monoid, G one of its subgroups, and e the identity of G. Let M denote the kernel of the homomorphism f: $\mathbf{Z}Te \otimes_G \mathbf{Z} \to \mathbf{Z}$ such that $t \otimes n \to n$ $(t \in T, n \in \mathbf{Z})$. Then for every T-module C we have an exact*

1980 *Mathematics Subject Classification* (1985 *Revision*). Primary 20M15, 20M50.

Translation of Vestnik Khar′kov. Univ. No. 221 (Prikl. Mat. i Mekh. vyp. (4) 46) (1981), 80–85; MR **84c**:20083.

sequence

$$\begin{aligned}&0 \to H^0(T,C) \to H^0(G,eC) \to \mathrm{Hom}_T(M,C) \to H^1(T,C)\\ &\quad\to \cdots \to H^n(T,C) \to H^n(G,eC) \to \mathrm{Ext}^n_T(M,C) \to H^{n+1}(T,C) \to \cdots.\end{aligned}$$

PROOF. A short exact sequence $0 \to M \to \mathbb{Z}Te \otimes_G \mathbb{Z} \to \mathbb{Z} \to 0$ generates a long exact sequence

$$\cdots \to H^n(T,C) \to \mathrm{Ext}^n_T(\mathbb{Z}Te \otimes_G \mathbb{Z}, C) \to \mathrm{Ext}^n_T(M,C) \to \cdots.$$

Since $\mathbb{Z}Te$ is a projective G-module (Lemma 2), by Lemma 2.1 of [1]

$$\mathrm{Ext}^n_T(\mathbb{Z}Te \otimes_G \mathbb{Z}, C) \cong H^n(G,eC),$$

q.e.d.

COROLLARY. *For a completely* 0*-simple semigroup* S *we have an exact sequence*

$$\begin{aligned}&0 \to H^0_0(S,A) \to H^0(G,eA) \to \mathrm{Hom}_{\overline{S}^1}(M,A) \to H^1_0(S,A)\\ &\quad\to \cdots \to H^n_0(S,A) \to H^n(G,eA) \to \mathrm{Ext}^n_{\overline{S}^1}(M,A) \to H^{n+1}_0(S,A) \to \cdots, \qquad (1)\end{aligned}$$

where $M = \mathrm{Ker}(\mathbb{Z}\overline{S}^1\langle e\rangle \otimes_G \mathbb{Z} \to \mathbb{Z})$.

The proof follows from Lemma 3 and Theorem 3 of [2].

4. To describe the term $\mathrm{Ext}^n_{\overline{S}^1}(M,A)$ of (1) we first investigate the structure of the $\overline{S}^1$-module M.

Let U denote the set of all elements of $\overline{S}$ of the form $\langle x^1,\dots,x^n,e_{i1}\rangle$ ($n \geq 0, i \in I$). Then each element of $\overline{S}^1\langle e\rangle$ ($e = e_{11}$) can be uniquely represented in the form $u\langle g_{11}\rangle$ ($u \in U, g \in G$), since

$$\langle x^1,\dots,x^n\rangle\langle e\rangle = \begin{cases}\langle x^1,\dots,x^{n-1},e_{i1}\rangle\langle (gp_{\lambda 1})_{11}\rangle, & \text{if } x^n = g_{i\lambda},\ p_{\lambda 1} \neq 0,\\ \langle x^1,\dots,x^{n-1},e\rangle, & \text{if } x^n = g_{i\lambda},\ p_{\lambda 1} = 0.\end{cases}$$

If $s \in \overline{S}$ is represented in the form $s = u\langle g_{11}\rangle$, we write $u = \omega(s)$.

It is easily seen that the group $\mathbb{Z}\overline{S}^1\langle e\rangle \otimes_G \mathbb{Z}$ is isomorphic to $\mathbb{Z}U$, and

$$M \cong \mathbb{Z}\{u - \langle e\rangle | u \in U\} \subseteq \mathbb{Z}U \qquad (2)$$

under this isomorphism, since $\sum_{u,g} m_{u,g}u\langle g_{11}\rangle \otimes n_{u,g} \in M$ if and only if

$$\sum_{u,g} m_{u,g}n_{u,g} = 0 \qquad (m_{u,g}, n_{u,g} \in \mathbb{Z}).$$

A structure of a unital $\overline{S}^1$-module is induced on $\mathbb{Z}U$: if $s \in \overline{S}^1$ and $u \in U$, then we define $s \circ u = \omega(su)$, where $\circ$ denotes the action of $\overline{S}^1$ on $\mathbb{Z}U$. Thus, if $s = \langle s^1,\dots,s^n\rangle \in \overline{S}$, then

$$s \circ u = \begin{cases}\langle s^1,\dots,s^{n-1},e_{i1}\rangle & \text{if } u = \langle e_{i1}\rangle,\ s^n = g_{i\lambda},\ p_{\lambda i} \neq 0,\\ su & \text{otherwise.}\end{cases}$$

Now we construct an irreducible system of generators of the $\overline{S}^1$-module M and defining relations of this system using the isomorphism (2). Let $B = \{\langle e_{i1}\rangle - \langle e\rangle | i \in I\backslash 1\}$ and $E(i) = \{\langle e_{i\lambda}\rangle | p_{\lambda i} \neq 0, p_{\lambda 1} \neq 0\}$.

LEMMA. *The set B is a system of generators of the $\overline{S}^1$-module M. An $\overline{S}^1$-linear combination*

$$\sum_{i\neq 1}\sum_{s\in\bar{S}^1} n_s^i s\circ(\langle e_{i1}\rangle - \langle e\rangle) \quad (n_s^i\in\mathbb{Z})$$

of elements of B equals zero if and only if $s\in\overline{S}^1E(i)$ for all $s\in\overline{S}^1$ and $i\in I\backslash 1$ such that $n_s^i\neq 0$. In particular, the system B of generators is irreducible.

PROOF. It follows from (2) that the module M is generated by elements of the form $u-\langle e\rangle$. Suppose that $u\in U$, $u=\langle x^1,\dots,x^{n-1},e_{i1}\rangle$ and $n\geq 2$. Then

$$u-\langle e\rangle = \langle x^1\rangle\circ(\langle x^2,\dots,x^{n-1},e_{i1}\rangle - \langle e\rangle) + (\langle x^1\rangle\circ\langle e\rangle - \langle e\rangle).$$

Therefore, if $|u|\geq 2$, then $u-\langle e\rangle$ can be expressed in terms of elements of lesser length, and hence the $\overline{S}$-module M is generated by B.

To prove the second part of the lemma, note that $\langle e_{1\lambda}\rangle\circ\langle e_{i\lambda}\rangle = \langle e_{1\lambda}\rangle\circ\langle e\rangle = \langle e\rangle$, if $\langle e_{1\lambda}\rangle\in E(i)$. Therefore $s\circ(\langle e_{i1}\rangle-\langle e\rangle)=0$ for all $s\in\overline{S}^1E(i)$.

Let an $\overline{S}^1$-linear combination of elements of B with nonzero coefficients $n_s^i\in\mathbb{Z}$ equal zero:

$$\sum_{i\neq 1}\sum_{s\in\bar{S}} n_s^i s\circ(\langle e_{i1}\rangle-\langle e\rangle) + \sum_{i\neq 1} n_1^i(\langle e_{i1}\rangle - \langle e\rangle) = 0. \tag{3}$$

By the previous remark we may assume that $n_s^i=0$ in (3) for $s\in\overline{S}^1E(i)$. If $n_s^i=0$ for all $s\in\overline{S}$ and $i\in I\backslash 1$, then (3) takes the form

$$\sum_{i\neq 1} n_1^i(\langle e_{i1}\rangle-\langle e\rangle)=0,$$

so that $n_1^i=0$ for all $i\in I\backslash 1$ thus contradicting the assumption about the coefficients of the linear combination. Let $t\in\overline{S}$ be an element such that $n_t^i\neq 0$ for some $i\in I\backslash 1$, and let the length $|t|=r$ be maximal. Since $t\notin\overline{S}^1E(i)$, we have either $|t\circ\langle e_{i1}\rangle|=r+1$ or $|t\circ\langle e\rangle|=r+1$. For example, let $|t\circ\langle e\rangle|=r+1$. Since $t\circ\langle e\rangle\neq t\circ\langle e_{i1}\rangle$ for $i\neq 1$, there exists an element s of $\overline{S}$, $s\neq t$, such that $n_s^j\neq 0$ and $s\circ\langle e_{j1}\rangle = t\circ\langle e\rangle$ for some $j\in I$. It follows from $|s|\leq r$ and $|s\circ\langle e_{j1}\rangle|=r+1$ that $|s|=r$. Let $s=\langle s^1,\dots,s^r\rangle$ and $t=\langle t^1,\dots,t^r\rangle$. It follows from $s\circ\langle e_{j1}\rangle = t\circ\langle e\rangle$ and $|t\circ\langle e\rangle|=r+1$ that $\langle s^1,\dots,s^r,e_{j1}\rangle=\langle t^1,\dots,t^r,e\rangle$, i.e. $s=t$ contrary to our assumption. This completes the proof.

5. As in [1], introduce the symbols ε_i ($i\in I\backslash 1$) and consider a free $\overline{S}^1$-module $\sum_{i\neq 1}\mathbb{Z}\overline{S}^1\varepsilon_i$ spanned by them. The correspondence $\varepsilon_i\to\langle e_{i1}\rangle-\langle e\rangle$ can be extended to an epimorphism $\sum_{i\neq 1}\mathbb{Z}\overline{S}^1\varepsilon_i\to M$. Lemma 4 can now be restated as follows.

PROPOSITION. *There exists an exact sequence*

$$0\to\sum_{i\neq 1}\mathbb{Z}\overline{S}^1E(i)\varepsilon_i\to\sum_{i\neq 1}\mathbb{Z}\overline{S}^1\varepsilon_i\to M\to 0$$

(if $E(i) = \varnothing$ for some i, we assume that $\mathbb{Z}\overline{S}^1 E(i) = 0$).

Considering the corresponding long exact sequence

$$\cdots \to \operatorname{Ext}^n_{\overline{S}^1}(M, A) \to \operatorname{Ext}^n_{\overline{S}^1}\left(\sum_{i\neq 1} \mathbb{Z}\overline{S}^1 \varepsilon_i, A\right)$$
$$\to \operatorname{Ext}^n_{\overline{S}^1}\left(\sum_{i\neq 1} \mathbb{Z}\overline{S}^1 E(i)\varepsilon_i, A\right) \to \cdots$$

and taking into account the projectivity of the $\overline{S}^1$-module $\sum_{i\neq 1} \mathbb{Z}\overline{S}^1 \varepsilon_i$ we obtain

COROLLARY. *For all $n \geq 2$*

$$\operatorname{Ext}^n_{\overline{S}^1}(M, A) \cong \operatorname{Ext}^{n-1}_{\overline{S}^1}\left(\sum_{i\neq 1} \mathbb{Z}\overline{S}^1 E(i)\varepsilon_i, A\right) \cong \prod_{i\neq 1} \operatorname{Ext}^{n-1}_{\overline{S}^1}(\mathbb{Z}\overline{S}^1 E(i), A)\varepsilon_i.$$

6. THEOREM. *The long exact sequence* (2) *is decomposable into the exact sequence*

$$\begin{aligned} 0 &\to H^0_0(S, A) \to H^0(G, eA) \to \operatorname{Hom}_{\overline{S}^1}(M, A) \to H^1_0(S, A) \\ &\to H^1(G, eA) \to \operatorname{Ext}^1_{\overline{S}^1}(M, A) \to H^2_0(S, A) \to H^2(G, eA) \to 0 \end{aligned} \tag{4}$$

and splitting short exact sequences

$$0 \to \prod_{i\neq 1} \operatorname{Ext}^{n-2}_{\overline{S}^1}(\mathbb{Z}\overline{S}^1 E(i), A)\varepsilon_i \to H^n_0(S, A) \to H^n(G, eA) \to 0 \tag{5}$$

for $n \geq 3$. Moreover, if $n \geq 3$ then

$$H^n_0(S, A) \cong H^n(G, eA).$$

PROOF. Let $X \to \mathbb{Z}$ be the standard G-projective resolvent of the G-module $\mathbb{Z}$ (see [4]), and let $Y \to M$ be the standard $\overline{S}^1$-projective resolvent of the $\overline{S}$-module M, where we can assume that $Y_0 = \sum_{i\neq 1} \mathbb{Z}\overline{S}^1 \varepsilon_i$. Then, by Lemma 2.1 of [1], $\mathbb{Z}\overline{S}^1\langle e\rangle \otimes_G X \to \mathbb{Z}\overline{S}^1\langle e\rangle \otimes_G \mathbb{Z}$ is a projective resolvent of the $\overline{S}^1$-module $\mathbb{Z}\overline{S}^1\langle e\rangle \otimes_G \mathbb{Z}$.

Next we define a homomorphism $\overline{f}\colon Y \to \mathbb{Z}\overline{S}^1\langle e\rangle \otimes_G X$ of complexes which is a mapping over the homomorphism $f\colon M \to \mathbb{Z}\overline{S}^1\langle e\rangle \otimes_G \mathbb{Z}$ (see [4]) so that $\overline{f}_n = 0$ for $n \geq 2$ (see the diagram):

$$\begin{array}{ccccccc} M & \xleftarrow{\varphi} & Y_0 & \xleftarrow{\partial_0} & Y_i & \xleftarrow{\partial_1} & \cdots \\ \downarrow f & & \downarrow \overline{f}_0 & & \downarrow \overline{f}_1 & & \\ \mathbb{Z}\overline{S}^1\langle e\rangle \otimes_G \mathbb{Z} & \xleftarrow{\psi} & \mathbb{Z}\overline{S}^1\langle e\rangle \otimes_G X_0 & \xleftarrow{\delta_0} & \mathbb{Z}\overline{S}^1\langle e\rangle \otimes_G X_1 & \xleftarrow{\delta_1} & \cdots \end{array}$$

As in [1], we define $\overline{f}_0(\varepsilon_i) = (\langle e_{i1}\rangle - \langle e\rangle) \otimes [\,]$, where $[\,]$ is a generating element of the module X_0. It is easy to see that $f_\varphi = \psi\overline{f}_0$. Further, suppose that $y \in Y_1$

and $\partial_0 y = \sum_{i\neq 1}\sum_{s\in \bar{S}^1} n_s^i s\varepsilon_i$. Since $\varphi\partial_0 = 0$, we have $\partial_0 Y_1 \subseteq \sum_{i\neq 1} \mathbb{Z}\overline{S}^1 E(i)\varepsilon_i$ (see Proposition 5). Therefore

$$\partial_0 y = \sum_{i\neq 1}\sum_{s\in \bar{S}^1 E(i)} n_s^i s\varepsilon_i = \sum_{i\neq 1}\sum_{s\in \bar{S}^1}\sum_{\langle e_{1\lambda}\rangle\in E(i)} n^i_{s\langle e_{1\lambda}\rangle} s\langle e_{1\lambda}\rangle\varepsilon_i.$$

However this decomposition is not uniquely determined since, for $s\neq t$, the equality $s\langle e_{1\lambda}\rangle = t\langle e_{1\lambda}\rangle$ is possible. The element $s\in\overline{S}$ is uniquely determined by the product $s\langle e_{1\lambda}\rangle$ if one demands that $s\in\overline{S}\langle e\rangle$. Therefore we may express $\partial_0 y$ in the form

$$\partial_0 y = \sum_{i\neq 1}\sum_{\langle e_{1\lambda}\rangle\in E(i)}\sum_{s\in\bar{S}\langle e\rangle} n^i_{s\langle e_{1\lambda}\rangle} s\langle e_{1\lambda}\rangle\varepsilon_i. \tag{6}$$

It follows that

$$\overline{f}_0\partial_0 y = \sum_{i\neq 1}\sum_{\langle e_{1\lambda}\rangle\in E(i)}\sum_{s\in\bar{S}\langle e\rangle} n^i_{s\langle e_{1\lambda}\rangle}(s(\langle p_{\lambda i}\rangle - \langle e\rangle)\otimes[\,] - s(\langle p_{\lambda 1}\rangle - \langle e\rangle\otimes[\,])$$
$$= \delta_0\sum_{i\neq 1}\sum_{\langle e_{1\lambda}\rangle\in E(i)}\sum_{s\in\bar{S}\langle e\rangle} n^i_{s\langle e_{1\lambda}\rangle} s\langle e\rangle\otimes([p_{\lambda i}] - [p_{\lambda 1}]).$$

Thus, if we assume that

$$\overline{f}_1 y = \sum_{i\neq 1}\sum_{\langle e_{1\lambda}\rangle\in E(i)}\sum_{s\in\bar{S}\langle e\rangle} n^i_{s\langle e_{1\lambda}\rangle} s\otimes([p_{\lambda i}] - [p_{\lambda 1}]),$$

where the coefficients $n^i_{s\langle e_{1\lambda}\rangle}$ are determined by (6), we obtain $\overline{f}_0\partial_0 = \delta_0\overline{f}_1$. Moreover, $\overline{f}_1\partial_1 = 0$, since if $z\in Y_2$ and $y = \partial_1 z$ then $\partial_0 y = 0$ and in (6) all coefficients $n_s^i = 0$; hence $\overline{f}_1\partial_1 z = \overline{f}_1 y = 0$.

Next we show that $\overline{f}_1$ is an $\overline{S}^1$-homomorphism. Suppose that $y\in Y_1$, $t\in S^1$, ∂_0 is defined by (6), and

$$\partial_0(ty) = \sum_{i\neq 1}\sum_{\langle e_{1\lambda}\rangle\in E(i)}\sum_{x\in\bar{S}\langle e\rangle} m^i_{x\langle e_{1\lambda}\rangle} x\langle e_{1\lambda}\rangle\varepsilon_i.$$

Comparing the coefficients in the equality $\partial_0(ty) = t\partial_0 y$, we obtain

$$m^i_{x\langle e_{1\lambda}\rangle} = \sum_{s:ts=x} n^i_{s\langle e_{1\lambda}\rangle}.$$

Therefore

$$\overline{f}_1(ty) = \sum_{i\neq 1}\sum_{\langle e_{1\lambda}\rangle\in E(i)}\sum_{x\in\bar{S}\langle e\rangle}\sum_{s:ts=x} n^i_{s\langle e_{1\lambda}\rangle} x\otimes([p_{\lambda i}] - [p_{\lambda 1}]) = t\overline{f}_1 y.$$

Thus, the mapping $\overline{f}$ has been constructed. Applying Lemma 1.1 of [1] to the exact sequence

$$0\to M\to \mathbb{Z}\overline{S}^1\langle e\rangle\otimes_G \mathbb{Z}\to\mathbb{Z}\to 0$$

and the mapping f, we now obtain the sequences (4) and (5).

To prove the last statement, transform the sandwich matrix P as follows: multiply the λth row by $p_{\lambda 1}^{-1}$ if $p_{\lambda 1}\neq 0$, and leave it intact otherwise. Applying this to every row, we arrive at a matrix p' in which the entries $p'_{\lambda 1}$ are either 0 or

1 for all $\lambda \in \Lambda$. It is known [3] that the semigroup $S^0(G; I, \Lambda; p')$ is isomorphic to S. Therefore we can assume that the entries $p_{\lambda 1}$ of P are either 0 or 1. In this case the elements $\langle e_{1\lambda} \rangle$ are idempotent for $p_{\lambda 1} = 1$, and the $\overline{S}^1$-module $\mathbb{Z}\overline{S}^1 E(i)$ is decomposable as a direct sum

$$\mathbb{Z}\overline{S}^1 E(i) = \sum_{\langle e_{1\lambda} \rangle \in E(i)} \mathbb{Z}\overline{S}^1 \langle e_{1\lambda} \rangle.$$

All summands of this sum are $\overline{S}^1$-projective, since $\langle e_{1\lambda} \rangle \in E(i)$ implies that $\langle e_{1\lambda} \rangle$ is idempotent. Therefore the module $\mathbb{Z}\overline{S}^1 E(i)$ is projective, and hence

$$\operatorname{Ext}^n_{\overline{S}^1}(\mathbb{Z}\overline{S}^1 E(i), A) = 0 \quad \text{for } n \geq 1.$$

It follows from (5) that $H_0^n(S, A) \cong H^n(G, eA)$ for $n \geq 3$, q.e.d.

7. Analogously to Corollary 3.5 of [1] we can deduce from the previous theorem a description of the groups $H_0^1(S, A)$ and $H_0^2(S, A)$ in the case when G is a trivial group.

COROLLARY. *If S is an inverse completely 0-simple semigroup then*

$$H_0^n(S, A) \cong H^n(G, eA)$$

for all $n \geq 2$. Furthermore, there exists an exact sequence

$$0 \to H_0^0(S, A) \to H^0(G, eA) \to \operatorname{Hom}_{\overline{S}^1}(M, A)$$
$$\to H_0^1(S, A) \to H^1(G, eA) \to 0.$$

PROOF. By Theorem 6 it suffices to establish the isomorphism for $n = 2$. Since $|I| = |\Lambda|$ and the sandwich matrix is the identity one for any inverse completely 0-simple semigroup [3], we obtain $E(i) = \varnothing$ for $i \neq 1$. Then $M \cong \sum_{i \neq 1} \mathbb{Z}\overline{S}^1 \varepsilon_i$ by Proposition 5, and hence M is a projective $\overline{S}^1$-module. Therefore $\operatorname{Ext}^1_{\overline{S}^1}(M, A) = 0$, and it follows from (4) that $H_0^2(S, A) \cong H^2(G, eA)$.

REFERENCES

1. William R. Nico, *On the cohomology of finite semigroups*, J. Algebra **11** (1969), 598–612.

2. B. V. Novikov, 0-*cohomology of semigroups*. Theoretical and Applied Questions of Differential Equations and Algebra (A. N. Sharkovskiĭ, editor), "Naukova Dumka", Kiev, 1978, pp. 185–188. (Russian)

3. A. H. Clifford and G. B. Preston, *The algebraic theory of semigroups*. Vol. 1, Amer. Math. Soc., Providence, R.I., 1961.

4. Henri Cartan and Samuel Eilenberg, *Homological algebra*, Princeton Univ. Press, Princeton, N.J., 1956.

Translated by B. M. SCHEIN

Amer. Math. Soc. Transl.
(2) Vol. **134**, 1987

Some Properties of d-Convex Functions. I

V. P. SOLTAN

The theory of convex sets in a linear space, and its applications, make extensive use of the concept of convex functions. Consequently it is to be expected that for each analog of linear convexity one can define a corresponding analog of convex functions. One of the natural analogs of convexity in a metric space is d-convexity, which has been introduced independently by several authors (see, for example, [1]–[4]). Let us recall that a set A in a metric space (X, d) is said to be d-convex if, for arbitrary points x, $y \in A$, the metric segment

$$\langle x, y\rangle = \{z \in X\colon d(x,z) + d(z,y) = d(x,y)\}$$

is contained in A. One possible version of the definition of metric convexity of a function was given in [5]. In the present paper we give detailed proofs and amplify the statements of some of the results announced in [5]. We note that an attempt at a definition of metrically convex functions was made in [4]. In what follows, we write $R^* = R \cup \{+\infty\}$.

DEFINITION. A function $f\colon X \to R^*$ is called *d-convex* if the inequality

$$f(z) \leq \frac{d(z,y)}{d(x,y)} f(x) + \frac{d(x,z)}{d(x,y)} f(y) \tag{1}$$

is satisfied for all points $x, y, z \in X$ $(x \neq y)$ that satisfy the equation $d(x,z) + d(z,y) = d(x,y)$.

THEOREM 1. *For a d-convex function $f\colon X \to R^*$ and every $\alpha \in R$ the level sets $\{z \in X\colon f(z) \leq \alpha\}$ and $\{z \in X\colon f(z) < \alpha\}$ are d-convex.*

1980 *Mathematics Subject Classification* (1985 *Revision*). Primary 26A51, 52A07, 54E35.

Translation of Izv. Akad. Nauk Moldav. SSR Ser. Fiz.-Tekhn. Mat. Nauk **1980**, no. 2, 27–31; MR **83a**:52004a.

PROOF. Let us show, for example, that the set $A = \{z \in X: f(z) \leq \alpha\}$ is d-convex. Let $x, y \in A$ and $z \in \langle x, y \rangle$. Then

$$f(z) \leq \frac{d(z,y)}{d(x,y)} f(x) + \frac{d(x,z)}{d(x,y)} f(y) \leq \left[\frac{d(z,y) + d(x,z)}{d(x,y)}\right] \cdot \alpha = \alpha.$$

Consequently $\langle x, y \rangle \subset A$. But this means that A is d-convex. This proves the theorem.

THEOREM 2. *The sum of two d-convex functions and the pointwise supremum of any family of d-convex functions are d-convex functions.*

PROOF. Let $x, y \in X$ $(x \neq y)$ and $z \in \langle x, y \rangle$.
If f_1 and f_2 are d-convex functions, then

$$(f_1 + f_2)(z) \leq \frac{d(z,y)}{d(x,y)}(f_1 + f_2)(x) + \frac{d(x,z)}{d(x,y)}(f_1 + f_2)(y).$$

If $\{f_\lambda\}$ is any family of d-convex functions, then

$$\sup_\lambda f_\lambda(z) \leq \frac{d(z,y)}{d(x,y)}\left[\sup_\lambda f_\lambda(x)\right] + \frac{d(x,z)}{d(x,y)}\left[\sup_\lambda f_\lambda(y)\right].$$

THEOREM 3. *If f is a d-convex function, and $\varphi: R \to R$ is a nondecreasing convex function, then $h = \varphi \circ f$ is d-convex.*

PROOF. Let $x, y \in X$ $(x \neq y)$ and $z \in \langle x, y \rangle$. Then

$$\begin{aligned} h(z) = \varphi(f(z)) &\leq \varphi\left(\frac{d(z,y)}{d(x,y)} f(x) + \frac{d(x,z)}{d(x,y)} f(y)\right) \\ &\leq \frac{d(z,y)}{d(x,y)} h(x) + \frac{d(x,z)}{d(x,y)} h(y). \end{aligned}$$

THEOREM 4. *Let $f_1, f_2, \ldots$ be a sequence of d-convex functions such that the sequence of numbers $f_1(z)$, $f_2(z), \ldots$ converges for each point $z \in X$. Then the function $f(x) = \lim_{k\to\infty} f_k(x)$ is d-convex.*

PROOF. If $z \in \langle x, y \rangle$, we have

$$\begin{aligned} f(z) = \lim_{k\to\infty} f_k(z) &\leq \lim_{k\to\infty}\left[\frac{d(z,y)}{d(x,y)} f_k(x) + \frac{d(x,z)}{d(x,y)} f_k(y)\right] \\ &= \frac{d(z,y)}{d(x,y)} f(x) + \frac{d(x,z)}{d(x,y)} f(y). \end{aligned}$$

THEOREM 5. *Let X be separable and let $f_1, f_2, \ldots$ be a sequence of d-convex continuous functions such that the sequence of numbers $f_1(z)$, $f_2(z), \ldots$ is bounded for each $z \in X$ (or even only for a dense subset of X). Then there exists a subsequence which converges in X to a d-convex function.*

PROOF. Let Z be a set, dense in X, such that the sequence $f_1(z)$, $f_2(z), \ldots$ is bounded for every $z \in Z$. Since X is separable, we can select from Z a countable set $Y = \{y_1, y_2, \ldots\}$ which is dense in X. Consider the double sequence $\{f_i(y_j)\}_{i,j=1}^\infty$ of numbers. Using Cantor's diagonal process, we can select a subsequence f_{i_1}, $f_{i_2}, \ldots$ of functions such that each sequence $f_{i_1}(y_j)$, $f_{i_2}(y_j), \ldots$

($j = 1, 2, \ldots$) converges. Since Y is dense in X, and the functions f_{i_k} are continuous, the sequence of numbers $f_{i_1}(x)$, $f_{i_2}(x), \ldots$ converges for every $x \in X$. By Theorem 4, the limit function $f(x) = \lim_{k\to\infty} f_{i_k}(x)$ is d-convex. The theorem is proved.

We say that a space X is *unbounded in every direction* if, for arbitrary points x, $y \in X$ and every number $K > 0$, there exists $z \in X$ such that $d(x, y) + d(y, z) \geq K$.

THEOREM 6. *If X is unbounded in every direction, then every d-convex function that is bounded above by some constant $\delta \in R$ is constant:* $f = \text{const}$.

PROOF. Suppose that there is a d-convex function f which is bounded above by a constant $\delta \in R$, but is not a constant. Then $f(x) \neq f(y)$ for some points x, $y \in X$. Let $f(x) < f(y)$. By hypothesis, for every positive integer n there is a point $z_n \in X$ such that $d(x, y) + d(y, z_n) = d(x, z_n) > n$. Set

$$\alpha_n = \frac{d(y, z_n)}{d(x, z_n)} f(x) + \frac{d(x, y)}{d(x, z_n)} f(z_n), \qquad n = 1, 2, \ldots.$$

Since $f(z_n) \leq \delta$, $d(x, z_n) \geq n$, and $d(y, z_n) = d(x, z_n) - d(x, y) < d(x, z_n)$, it is evident that $f(x) > \alpha_n$ for sufficiently large n. At the same time, it follows from (1) that $f(y) \leq \alpha_n$ for points $y \in \langle x, z_n \rangle$, which contradicts the hypothesis $f(x) < f(y)$. This contradiction establishes the theorem.

THEOREM 7. *Let $f \colon X \to R^*$ be a d-convex function, and let $z \in X$. Then, for every $y \in X$ and every sequence $z_k \in \langle z, y \rangle$ that converges to z, the limit $\lim_{k\to\infty} f(z_k)$, if it exists, does not exceed $f(z)$. If, moreover, this point is a passing point* (*see* [10]), *the limit* $\lim_{k\to\infty} f(z_k)$ *exists and equals* $f(z)$.

PROOF. It follows from the inequality

$$f(z_k) \leq \frac{d(z_k, y)}{d(x, y)} f(z) + \frac{d(z, z_k)}{d(z, y)} f(y)$$

that for every $\varepsilon > 0$ there is an integer N such that $f(z_k) < f(z) + \varepsilon$ for $k \geq N$. If $\lim_{k\to\infty} f(z_k)$ exists, then it is evident that $\lim f(z_k) \leq f(z) + \varepsilon$. Since $\varepsilon > 0$ can be chosen arbitrarily, we have $\lim f(z_k) \leq f(z)$.

If z is passable then $z \in \langle x, y \rangle$ for some point $x \in X$ different from z. It is easy to verify [6] that $z \in \langle z_k, x \rangle$. It follows from the inequality

$$f(z) \leq \frac{d(z, x)}{d(z_k, x)} f(z_k) + \frac{d(z_k, z)}{d(z_k, x)} f(x)$$

that for every $\delta > 0$ there is an integer M satisfying $f(z) < f(z_k) + \delta$ for $k \geq M$. From this and what was mentioned above we conclude that $\lim_{k\to\infty} f(z_k)$ exists and equals $f(z)$. The theorem is established.

We recall [7] that a function $f \colon X \to R^*$ is called *closed* (or *lower semicontinuous*) if $f(z) = \underline{\lim}_{y\to z} f(y)$ for every $z \in X$.

THEOREM 8. *A function $f\colon X \to R^*$ is closed if and only if the level sets $\{z \in X\colon f(z) \le \alpha\}$ are closed for all $\alpha \in R$.*

PROOF. Let f be closed, $\alpha \in R$ and $A = \{z \in X\colon f(z) \le \alpha\}$. If $x \in \overline{A}$, there is a sequence $\{y_k\} \subset A$ such that $y_k \to x$. Since f is closed, we have $f(x) \le \underline{\lim}_{k\to\infty} f(y_k) \le \alpha$. Consequently $x \in A$, and hence A is closed.

Conversely, for some function f let the sets $\{z \in X\colon f(z) \le \alpha\}$ be closed for all $\alpha \in R$, and let x be an arbitrary point of X. Let us show that $f(x) = \underline{\lim}_{y\to x} f(y)$. Suppose the contrary: let $\beta = \underline{\lim}_{y\to x} f(y) < f(x)$. Choose a number α such that $\beta < \alpha < f(x)$, and put $A = \{z \in X\colon f(z) \le \alpha\}$. It is easily seen that $x \in \overline{A}\backslash A$. Consequently A is not closed. This contradiction shows that $f(x) = \underline{\lim}_{y\to x} f(y)$. The theorem is established.

A metric space X is said to be *weakly Λ-convex*, where $\Lambda \subset [0,1]$, if, for all points x, $y \in X$, all $z \in \langle x, y\rangle$, and every number $\lambda \in \Lambda$, there exists $u \in X$ such that

$$d(x,u) = \lambda d(x,y) \quad \text{and} \quad d(u,y) = (1-\lambda)d(x,y),$$

and one of the relations $z \in \langle x, y\rangle$ or $z \in \langle u, y\rangle$ is satisfied.

From now on, we shall suppose that $\Lambda \cap (0,1) \ne \varnothing$.

We shall need the following lemma.

LEMMA 1. *If a function f satisfies (1) for the three points $z \in \langle x, y\rangle$ and $u \in \langle x, z\rangle$ then it satisfies (1) for the three points $u \in \langle x, y\rangle$.*

PROOF. That $u \in \langle x, y\rangle$ was proved in [6] (and independently in [7]):

$$\begin{aligned} f(u) &\le \frac{d(u,z)}{d(x,z)} f(x) + \frac{d(x,u)}{d(x,z)} f(z) \\ &\le \left(1 - \frac{d(x,u)}{d(x,z)}\right) f(x) + \frac{d(x,u)}{d(x,z)} \left[\left(1 - \frac{d(x,z)}{d(x,y)} f(x) + \frac{d(x,z)}{d(x,y)} f(y)\right)\right] \\ &= \frac{d(u,y)}{d(x,y)} f(x) + \frac{d(x,u)}{d(x,y)} f(y). \end{aligned}$$

We shall say that a function $f\colon X \to R^*$ is *d_λ-convex*, where $\lambda \in (0,1)$ is a given number, if $f(z) \le (1-\lambda)f(x) + \lambda f(y)$ for all x, y, $z \in X$ that satisfy $d(x,z) = \lambda d(x,y)$ and $d(z,y) = (1-\lambda)d(x,y)$. It is easy to construct an example of a d_λ-convex function that is not d-convex.

THEOREM 9. *Let X be a weakly Λ-convex space. A closed function $f\colon X \to R^*$ is d-convex if and only if it is d_λ-convex for each $\lambda \in \Lambda$.*

PROOF. If the points $x, y, z \in X$ satisfy the conditions $d(x,z) = \lambda d(x,y)$ and $d(z,y) = (1-\lambda)d(x,y)$ then $d(x,z) + d(z,y) = d(x,y)$. It follows that a d-convex function is d_λ-convex for each $\lambda \in \Lambda$.

Let us show that a function that satisfies the hypotheses of the theorem is d-convex. Let x, $y \in X$ $(x \ne y)$ and $z \in \langle x, y\rangle$. We select an infinite sequence of points z_j^i of X $(j = 1, 2, \dots; i = 1, 2, 3)$ in the following way: $z_1^1 = x$, $z_1^3 = y$, $d(z_1^1, z_1^2) = \lambda d(z_1^1, z_1^3)$, $d(z_1^2, z_1^3) = (1-\lambda)d(z_1^1, z_1^3)$, and one of the following inclusions is satisfied: $z \in \langle z_1^1, z_1^2\rangle$ or $z \in \langle z_1^2, z_1^3\rangle$. If z_k^1, z_k^2, and z_k^3 have been selected, we select the one of $\langle z_k^1, z_k^2\rangle$ or $\langle z_k^2, z_k^3\rangle$ that contains z, for example $\langle z_k^1, z_k^2\rangle$, put

$$z_{k+1}^1 = z_k^1, \qquad z_{k+1}^3 = z_k^3, \qquad d(z_{k+1}^1, z_{k+1}^2) = \lambda d(z_{k+1}^1, z_{k+1}^3),$$
$$d(z_{k+1}^2, z_{k+1}^3) = (1-\lambda) d(z_{k+1}^1, z_{k+1}^3),$$

and then one of the inclusions $z \in \langle z_{k+1}^1, z_{k+1}^2\rangle$ or $z \in \langle z_{k+1}^2, z_{k+1}^3\rangle$ is satisfied.

We can easily show, by using Lemma 1 and induction on k, that

$$f(z_k^2) \le \frac{d(z_k^2, y)}{d(x,y)} f(x) + \frac{d(x, z_k^2)}{d(x,y)} f(y).$$

Since $d(z, z_k^2) \le \delta^k d(x,y)$, where $\delta = \max\{\lambda, 1-\lambda\}$, we have $z_k^2 \to z$. Since f is closed, we have

$$\begin{aligned} f(z) &\le \varliminf_{k\to\infty} f(z_k^2) \\ &\le \varliminf_{k\to\infty} \left[\frac{d(z_k^2, y)}{d(x,y)} f(x) + \frac{d(x, z_k^2)}{d(x,y)} f(y)\right] \\ &= \frac{d(x,z)}{d(x,y)} f(x) + \frac{d(x,z)}{d(x,y)} f(y). \end{aligned}$$

Consequently f is d-convex. This completes the proof of the theorem.

Let us recall [6] that a metric space X is called *Λ-convex*, where $\Lambda \subset [0,1]$, if, for every x, $y \in X$ and every $\lambda \in \Lambda$, there is a point $z \in X$ such that $d(x,z) = \lambda d(x,y)$ and $d(z,y) = (1-\lambda)d(x,y)$. The following theorem is proved like Theorem 2 of [4].

THEOREM 10. *Let the space X be Λ-convex and let $\lambda \in \Lambda$. If a d_λ-convex function $f\colon X \to R^*$ has a local minimum on the d-convex set $A \subset X$, this minimum is a global minimum on A.*

PROOF. Let $U_r(z)$ be the closed ball of radius r with center at z. If x is a local minimum point for f on A, then there is a number $r > 0$ such that $f(y) \ge f(x)$ for $y \in A \cap U_r(x)$. Suppose that $f(x)$ is not a global minimum for f on A. Then $f(x) > f(z)$ for some $z \in A$. Let $z_1, z_2, \dots$ be a sequence of points in X such that $z_1 = z$, $d(x, z_k) = \lambda d(x, z_{k-1})$, and $d(z_k, z_{k-1}) = (1-\lambda)d(x, z_{k-1})$, $k = 2, 3, \dots$. Since $z_k \in \langle x, z_{k-1}\rangle$, the d-convexity of A implies that $z_1, z_2, \dots$ are contained in A. It follows from the inequality $d(x, z_k) \le \lambda^k d(x,z)$ that $z_k \to x$. Consequently $z_k \in A \cap U_r(x)$ for sufficiently large k.

Using Lemma 1, we obtain, by induction on k,

$$f(z_k) \le \frac{d(z_k, z)}{d(x,z)} f(x) + \frac{d(x, z_k)}{d(x,z)} f(z).$$

Since $d(x, z_k) \to 0$ and $d(z_k, x) < d(x,z)$, for sufficiently large k the right-hand side of the preceding inequality is less than $f(x)$. Consequently there is a point $z_k \in A \cap U_r(x)$ for which $f(z_k) < f(x)$, and this contradicts the fact that f has a local minimum at x. This contradiction establishes the theorem.

We say that a function $f\colon X \to R^*$ is *nonrecessive* [8] if the level sets $\{z \in X\colon f(z) \leq \alpha\}$ are bounded for all $\alpha \in R$. We recall [9] that a space X is called finitely compact if every bounded subset of X is precompact.

THEOREM 11. *Let the complete space X be finitely compact and let the closed function $f\colon X \to R^*$ be nonrecessive. Then its infimum is attained in X. Moreover, every sequence of points $z_1, z_2, \ldots$ in X that satisfies $\inf_k f(z_k) = \inf_k f$ is bounded, and all its limit points belong to the set $\{z \in X\colon f(z) = \inf f\}$.*

PROOF. Set $\alpha = \inf f$; we show that $A = \{z \in X\colon f(z) = \alpha\}$ is nonempty and compact. In fact, A can be represented in the form $A = \bigcap_{\beta > \alpha} A_\beta$, where $A_\beta = \{z \in X\colon f(z) \leq \beta\}$. By Theorem 8, each set A_β is closed. The function is nonrecessive, and therefore A_β is bounded. As the intersection of a decreasing sequence of compact sets A_β, the set A is nonempty and compact.

Let $z_1, z_2, \ldots$ be a sequence of points of X such that $\inf_k f(z_k) = \alpha$. Then for every $\beta > \alpha$ there is an integer N such that $f(z_k) \leq \beta$ for $k \geq N$. In other words, $z_k \in A_\beta$ for $k \geq N$. Since A_β is compact, the sequence $z_1, z_2, \ldots$ is bounded and has limit points. If z is a limit point of this sequence, the fact that f is closed implies that

$$f(z) \leq \varliminf_{k \to \infty} f(z_k) = \alpha.$$

Consequently $f(z) = \alpha$, and hence $z \in A$. This completes the proof.

REFERENCES

1. J. de Groot, *Some special metrics in general topology*, Colloq. Math. **6** (1958), 283–286.

2. A. D. Aleksandrov and V. A. Zalgaller, *Two-dimensional manifolds of bounded curvature*, Trudy Mat. Inst. Steklov. **63** (1962); English transl., *Intrinsic geometry of surfaces*, Amer. Math. Soc., Providence, R. I., 1967.

3. P. S. Soltan and K. F. Prisakaru, *Steiner's problem on graphs*, Dokl. Akad. Nauk. SSSR **198** (1971), 46–49; English transl. in Soviet Math. Dokl. **12** (1971).

4. T. T. Arkhipova and Ī. V. Sergiēnko, *On the formalization and solution of some problems of organizing the computing process in data processing systems*, Kibernetika (Kiev) **1973**, no. 5, 11–18; English transl. in Cybernetics **9** (1973).

5. V. P. Soltan and P. S. Soltan, *d-convex functions*, Dokl. Akad. Nauk SSSR **249** (1979), 555–558; English transl. in Soviet Math. Dokl. **20** (1979).

6. James S. W. Wong, *Remarks on metric spaces*, Nederl. Akad. Wetensch. Proc. Soc. A **69** = Indag. Math. **28** (1966), 70–73.

7. V. P. Soltan, *d-convex sets in finite metric spaces*, Izv. Akad. Nauk Moldav. SSR Ser. Fiz.-Tekhn. Mat. Nauk **1978**, no. 2, 29–33. (Russian)

8. R. Tyrrell Rockafellar, *Convex analysis*, Princeton Univ. Press, Princeton, N.J., 1970.

9. Herbert Busemann, *The geometry of geodesics*, Academic Press, 1955.

10. Leonard M. Blumenthal, *Theory and applications of distance geometry*, Clarendon Press, Oxford, 1953.

Translated by RALPH P. BOAS

Amer. Math. Soc. Transl.
(2) Vol. **134**, 1987

Some Properties of d-Convex Functions. II

V. P. SOLTAN

This paper is a continuation of [1]. We note that part of the results presented here were announced in [2]. In what follows, (X, d) always denotes a metric space X with metric d. For the convenience of the reader we recall that a function $f\colon X \to R^* = R \cup \{+\infty\}$ is called *d-convex* if for all points x, y, $z \in X$ $(x \neq y)$ that satisfy the equation $d(x,z) + d(z,y) = d(x,y)$, the inequality

$$f(z) \leq \frac{d(z,y)}{d(x,y)} f(x) + \frac{d(x,z)}{d(x,y)} f(y) \tag{1}$$

is satisfied.

We also recall that a set $A \subset X$ is called *d-convex* if, for all $x, y \in A$, the metric segment $\langle x, y\rangle = \{z \in X\colon d(x,z) + d(z,y) = d(x,y)\}$ belongs to A.

THEOREM 1. *For every d-convex function $f\colon X \to R^*$ the set* $\operatorname{Dom} f = \{z \in X\colon f(z) < \infty\}$ *is d-convex.*

PROOF. Consider the sets $A_n = \{z \in X\colon f(z) \leq n\}$, $n = 1, 2, \ldots$. It is easily seen that $\operatorname{Dom} f = \bigcup_{n \geq 1} A_n$ and $A_n \subset A_{n+1}$. Since each set A_n is d-convex [1], [2], and the union of an increasing sequence of d-convex sets is d-convex, the set $\operatorname{Dom} f$ is d-convex. This establishes the theorem.

We recall that a metric space X is called *convex* (or *convex in the sense of Menger*) if, for every pair of distinct points $x, y \in X$, there exists a point z, distinct from x and y, satisfying $d(x,z) + d(z,y) = d(x,y)$ [3].

THEOREM 2. *Let X be a complete convex metric space and let $f\colon X \to R^*$ be a continuous function. Then either* $\operatorname{Dom} f = X$ *or* $\operatorname{Dom} f = \varnothing$ *(i.e., $f(x) = \infty$ for $x \in X$).*

1980 *Mathematics Subject Classification* (1985 *Revision*). Primary 26A51, 52A07, 54E35, 46B20.

Translation of Izv. Akad. Nauk Moldav. SSR Ser. Fiz.-Tekhn. Mat. Nauk **1981**, no. 1, 21–26; MR **83a**: 52004b.

The proof follows from the definition and the fact that every two points of a complete convex metric space can be joined by a continuous path [3].

Let us consider the connection between the d-convexity of functions and their convexity in the ordinary sense.

THEOREM 3. *In every normed linear space E with metric $d(x,y) = \|x-y\|$, every d-convex function f is convex.*

PROOF. Let f be a d-convex function on E, and let $z \in [x,y]$. Then $z = \alpha x + (1-\alpha)y$ for some number α, $0 \le \alpha \le 1$. It is easily seen that $d(x,z) = (1-\alpha)d(x,y)$ and $d(z,y) = \alpha d(x,y)$. Consequently $d(x,z) + d(z,y) = d(x,y)$ and hence, by (1), $f(z) \le \alpha f(x) + (1-\alpha) f(y)$. But this means that f is convex. This completes the proof.

Recall that a space E is called *strictly normed* if for arbitrary vectors $e_1, e_2 \in E$ ($e_1 \ne 0$) the equation $\|e_1\| + \|e_2\| = \|e_1 + e_2\|$ implies that $e_2 = \lambda e_1$ for some $\lambda \ge 0$.

THEOREM 4. *The following conditions are equivalent.*

1) *Every convex function $f: E \to R^*$ is d-convex.*

2) *E is strongly normed.*

PROOF. If E is strictly normed, arbitrary points $x, y, z \in E$ for which $d(x,z) + d(z,y) = d(x,y)$ satisfy $z \in [x,y]$ (see, for example, [4]). From this it evidently follows that every convex function f on E is d-convex.

Conversely, let every convex function $f: E \to R^*$ be d-convex and let $A \subset E$ be any set. Put

$$f(x) = \begin{cases} 0 & \text{if } x \in A, \\ \infty & \text{if } x \notin A. \end{cases}$$

Since f is convex, it is, by hypothesis, d-convex. Then the set $A = \{z \in E: f(z) \le 0\}$ is also d-convex. Since $A \subset E$ is arbitrary, we conclude that E is strictly normed. This completes the proof of the theorem.

Recall that the closure of a function $f: X \to R^*$ is the function $\overline{f}$ defined by $\overline{f}(z) = \underline{\lim}_{y \to x} f(y)$ for each $z \in X$. It is known (see, for example, [5]) that the closure of a convex function in an n-dimensional linear space E^n is also a convex function. We have the following theorem.

THEOREM 5. *The closure of a d-convex function in a normed space E^n is also a d-convex function.*

PROOF. Let $f: E^n \to R^*$ be a d-convex function. It is easily seen that we need only verify the convexity of $\overline{f}$ on the set $\operatorname{Dom} \overline{f} = \{z \in E^n: \overline{f}(z) < \infty\}$. Let $x, y \in \operatorname{Dom} \overline{f}$ and $z \in \langle x, y \rangle$. Since $\operatorname{Dom} \overline{f}$ is d-convex (Theorem 1), we have $z \in \operatorname{Dom} \overline{f}$. It evidently follows from the definition of the closure of a function that $\operatorname{Dom} \overline{f} \subset \overline{\operatorname{Dom} f}$. Choose any point $w \in \operatorname{rint} \operatorname{Dom} f$ and put $x_\lambda = \lambda x + (1-\lambda) w$, $y_\lambda = \lambda y + (1-\lambda) w$, and $z_\lambda = \lambda z + (1-\lambda) w$ for $0 \le \lambda \le 1$. It is known (see, for example, [5]) that the half-open intervals $[w, x)$, $[w, y)$, and

$[w, z)$ belong to the set $\operatorname{rint}\operatorname{Dom} f$. Consequently $x_\lambda, y_\lambda, z_\lambda \in \operatorname{rint}\operatorname{Dom} f$ for $0 \leq \lambda < 1$. Since

$$\begin{aligned} d(x_\lambda, z_\lambda) + d(z_\lambda, y_\lambda) &= \|x_\lambda - z_\lambda\| + \|z_\lambda - y_\lambda\| = \lambda(\|x - z\| + \|z - y\|) \\ &= \lambda\|x - y\| = d(x_\lambda, y_\lambda), \end{aligned}$$

we have $z_\lambda \in \langle x_\lambda, y_\lambda \rangle$. The function f is d-convex, and therefore

$$f(z_\lambda) \leq \frac{d(z_\lambda, y_\lambda)}{d(x_\lambda, y_\lambda)} f(x_\lambda) + \frac{d(x_\lambda, z_\lambda)}{d(x_\lambda, y_\lambda)} f(y_\lambda).$$

By Theorem 3, the function f is convex. Every convex function φ on E^n is continuous on $\operatorname{rint}\operatorname{Dom}\varphi$ [5]. Consequently $\overline{f}(u) = f(u)$ for $u \in \operatorname{rint}\operatorname{Dom} f$, and therefore

$$\overline{f}(x) = \lim_{\lambda \to 1} f(x_\lambda), \quad \overline{f}(y) = \lim_{\lambda \to 1} f(y_\lambda), \quad \overline{f}(z) = \lim_{\lambda \to 1} f(z_\lambda).$$

Then

$$\begin{aligned} \overline{f}(z) = \lim_{\lambda \to 1} f(z_\lambda) &\leq \lim_{\lambda \to 1} \left[\frac{d(z_\lambda, y_\lambda)}{d(x_\lambda, y_\lambda)} f(x_\lambda) + \frac{d(x_\lambda, z_\lambda)}{d(x_\lambda, y_\lambda)} f(y_\lambda) \right] \\ &= \frac{d(z, y)}{d(x, y)} \overline{f}(x) + \frac{d(x, z)}{d(x, y)} \overline{f}(y). \end{aligned}$$

Consequently f is d-convex. This completes the proof of the theorem.

We note that the following analog of Theorem 3 is valid for every normed linear space E.

THEOREM 6. *Let the d-convex function $f: E \to R^*$ have the property that the set* $\operatorname{Dom} f$ *has nonempty interior and f is bounded in a neighborhood of some point $x \in \operatorname{int}\operatorname{Dom} f$. Then the closure of f is also d-convex.*

The proof is similar to the proof of Theorem 5. It depends on the fact that the boundedness of a convex function f in the neighborhood of a point $x \in \operatorname{int}\operatorname{Dom} f$ implies the continuity of f [6].

It is easy to construct examples showing that Theorems 5 and 6 fail in arbitrary metric spaces.

THEOREM 7. *If the closure of every d-convex function $f: X \to R^*$ is d-convex, the closure of every d-convex set $A \subset X$ is also d-convex.*

PROOF. Let $A \subset X$ be a d-convex set. Define a function $f: X \to R^*$ as follows:

$$f(x) = \begin{cases} 0 & \text{for } x \in A, \\ \infty & \text{for } x \notin A. \end{cases}$$

It is easily verified that f is d-convex. Then, by hypothesis, its closure $\overline{f}$ is also d-convex. Since $\overline{A} = \{z \in X: \overline{f}(z) = 0\}$, the set $\overline{A}$ is d-convex. This completes the proof.

In the remainder of this paper we consider a circle of problems connected with a metric analog of the Kreĭn-Mil′man theorem.

For every set $A \subset X$ the *d-convex cover* d-conv A is defined in a natural way as the smallest d-convex subset of X that contains A. Since every space X is d-convex and the intersection of any collection of d-convex sets is d-convex, the set d-conv A exists and is uniquely defined.

THEOREM 8. *Let $f: X \to R^*$ be a d-convex function and let $A \subset X$. If $f(z) \leq \alpha$ for $z \in A$, then $f(z) \leq \alpha$ for $z \in d\text{-conv}\, A$.*

PROOF. Consider the set $B = \{z \in X: f(z) \leq \alpha\}$. Since it is d-convex and $A \subset B$, we have $d\text{-conv}\, A \subset d\text{-conv}\, B = B$. Consequently $f(z) \leq \alpha$ for $z \in d\text{-conv}\, A$. This complees the proof.

THEOREM 9. *Let F be the family of d-convex functions on X. Then for every set $A \subset X$*

$$d\text{-conv}\, A = \bigcap \left\{ z \in X: f(z) \leq \sup_{x \in A} f(x),\ f \in F \right\}. \tag{2}$$

PROOF. Let $f \in F$ and $\alpha = \sup_{x \in A} f(x)$. Then by Theorem 8 we have $d\text{-conv}\, A \subset \{z \in X: f(z) \leq \alpha\}$. Consequently

$$d\text{-conv}\, A \subset \left\{ z \in X: f(z) \leq \sup_{x \in A} f(x), f \in F \right\}.$$

To prove (2) we consider the function

$$\varphi(x) = \begin{cases} 0 & \text{if } x \in d\text{-conv}\, A, \\ \infty & \text{if } x \notin d\text{-conv}\, A. \end{cases}$$

It is easily seen that $\varphi \in F$. Since $d\text{-conv}\, A = \{z \in X: \varphi(z) \leq 0\}$, equation (2) is established.

DEFINITION 1. A *d-face* of the set $A \subset X$ is any subset B of A that satisfies the following condition: if $z \in B$ and $z \in \langle x, y \rangle$ for some points x and y of A, different from z, then $x, y \in B$. A d-face B of A is called *proper* if $B \neq A$. If B consists of the single point z, we call z a *d-extreme point* of A.

We note that some properties of d-extreme points are studied in [7] and [8]. The following properties of d-faces are easily established.

1°. *Every union or intersection of d-faces of the set $A \subset X$ is a d-face of A.*

2°. *If A is a d-face of B and B is a d-face of C, then A is a d-face of C.*

3°. *If $A \subset B \subset C$ and A is a d-face of C, then A is a d-face of B.*

4°. *Let G be a family of subsets of X such that, for every pair of elements of G, one is a d-face of the other. Then $B = \bigcap\{A \in G\}$ (if not empty) is a d-face of every set $A \in G$.*

We recall [3] that a point $x \in A$ is a *passing point* in A if, for every point $y \in A$, there is a point z, different from x, satisfying $x \in \langle y, z \rangle$.

It evidently follows from the definition that every proper d-face of A fails to contain passing points in A.

THEOREM 10. *Let $f: X \to R^*$ be a d-convex function and $A \subset X$. Then the set $\{z \in A: f(z) = \sup f(x): x \in A\}$ (if not empty) is a d-face of A.*

PROOF. Set $\alpha = \sup_{x\in A} f(x)$ and $B = \{z \in A: f(z) = \alpha\}$. Let $z \in B$ and $x \in \langle x, y\rangle$ for some x and y in A, different from z. Then since f is d-convex, (1) is satisfied. Since $\max\{f(x), f(y)\} \leq f(z)$ and neither of the numbers $d(z, y)$ or $d(x, z)$ is zero, this inequality is possible only when $f(x) = f(y) = f(z)$. Consequently $x, y \in B$. But this means that B is a d-face of A. This completes the proof.

The following theorem follows from Theorem 10.

THEOREM 11. *If the d-convex function $f: X \to R^*$ attains a maximum on the set A at a passing point of A, the function f is constant on A.*

PROOF. Let $\alpha = \sup_{x\in A} f(x)$ and $B = \{z \in A: f(z) = \alpha\}$. By Theorem 8, the set B is a d-face of A. It follows from the remark preceding Theorem 8 that the d-face B is not proper in A, i.e. $B = A$. Consequently $f(x) = \alpha$ for every $x \in A$, i.e. f is constant on A. This completes the proof.

DEFINITION 2. We say that a metric space X has the *d-separation property* if, for every d-convex set $A \subset X$ and every point $z \notin A$, there is a continuous d-convex function $f: X \to R^*$ that satisfies the inequality $f(z) > \sup\{f(x): x \in A\}$.

Let Φ be the set of continuous d-convex functions $f: X \to R^*$.

THEOREM 12. *If X has the d-separation property, every set $A \subset X$ has the property that*

$$\overline{d\text{-conv}\, A} = \bigcap \left\{ z \in X: f(z) \leq \sup_{x\in A} f(x),\ f \in \Phi \right\}.$$

PROOF. Let

$$B = \bigcap \left\{ z \in X: f(z) \leq \sup_{x\in A} f(x),\ f \in \Phi \right\}.$$

We can show as in Theorem 9 that $d\text{-conv}\, A \subset B$. It is known [1], [2] that, for every continuous function $f: X \to R^*$ and for every number $\alpha \in R$, the set $\{z \in X: f(z) \leq \alpha\}$ is closed. Consequently B, as an intersection of closed sets, is closed. Hence $\overline{d\text{-conv}\, A} \subset B$. Let us suppose that $\overline{d\text{-conv}\, A} \neq B$. Then there is a point $x \in B \setminus \overline{d\text{-conv}\, A}$. Since X has the d-separation property, there is a continuous function $f: X \to R^*$ such that $f(z) > \sup\{f(x): x \in A\}$. Then $z \notin B$. This contradiction shows that $\overline{d\text{-conv}\, A} = B$, and the proof is complete.

COROLLARY 1. *If the space X has the d-separation property, the closure of every d-convex subset of X is also d-convex.*

THEOREM 13. *If the complete space X has the d-separation property, every nonempty compact subset of X has at least one d-extreme point.*

PROOF. Let A be a nonempty compact subset of X. If A is a singleton, this point is a d-extreme point of A. Let x and y be distinct points of A. Since the sets $\{x\}$ and $\{y\}$ are d-convex, there is a continuous d-convex function $f: X \to R^*$ such that $f(x) > f(y)$. Consider the set

$$B = \left\{ z \in A: f(z) = \sup_{x \in A} f(x) \right\}.$$

Since A is compact and f is continuous, B is not empty. By Theorem 10, the set B is a d-face of A. If B is a singleton, it is a d-extreme point of A. If B contains more than one point, we apply the procedure described above to B.

Using property 4° of d-faces and remembering that in a complete metric space the intersection of any number of compact sets is not empty, we see that A has a d-extreme point. This completes the proof.

THEOREM 14. *If the complete space X has the d-separation property, every compact d-convex subset of X is the closure of the d-convex cover of its d-extreme points.*

PROOF. Let A be any nonempty compact d-convex set and let B be the set of its d-extreme points. By Theorem 13, B is not empty. Since $B \subset A$ we have $\overline{d\text{-conv}\, B} \subset A$. Let us show that $A \subset \overline{d\text{-conv}\, B}$. Suppose the contrary. Then there is a point $z \in A \backslash d\text{-conv}\, B$. By hypothesis, X has the d-separation property. Consequently there is a function $f \in \Phi$ such that

$$f(z) > \sup\{f(x): x \in \overline{d\text{-conv}\, B}\}.$$

Let $C = \{z \in A: f(z) = \sup_{x \in A} f(x)\}$. Evidently C is not empty and $C \cap \overline{d\text{-conv}\, B} = \varnothing$. By Theorem 13, C has at least one d-extreme point, which is also d-extreme for A (property 2° of d-faces). But then $C \cap B \neq \varnothing$, contradicting the equation $C \cap \overline{d\text{-conv}\, B} = \varnothing$. This contradiction establishes that $A \subset \overline{d\text{-conv}\, B}$. Consequently $A = \overline{d\text{-conv}\, B}$. This completes the proof.

We see that the proof of Theorem 14 is analogous to the proof of the Kreĭn-Mil′man theorem. This suggests the problem of comparing these theorems for the case of a normed linear space. The following result shows that, at least in a finite-dimensional normed space E^n, the two theorems are essentially the same.

THEOREM 15. *If the normed space E^n with metric $d(x, y) = \|x - y\|$ has the d-separation property, it is strictly normed.*

PROOF. Let $x, y \in E^n$ be distinct points. Since the sets $\{x\}$ and $\{y\}$ are d-convex, there is a continuous d-convex function $f: E^n \to R^*$ such that $f(x) > f(y)$. Let $f(x) > \alpha > f(y)$ and let $A = \{z \in E^n: f(z) \leq \alpha\}$. Since f is continuous and not a constant, the d-convex set A is closed and has a nonempty interior $\operatorname{int} A \neq \varnothing$. It follows from the results of [9] that there is a d-convex hyperplane $H \subset E^n$ that strictly separates x from A. Consequently H strictly separates x and y. Since x and y were arbitrary, it follows from [9] that E is strictly normed. This completes the proof.

It remains only to remark that in a strictly normed space the concepts of convexity and d-convexity coincide.

REFERENCES

1. V. P. Soltan, *Some properties of d-convex functions.* I, Izv. Akad. Nauk Moldav. SSR Ser. Fiz.-Tekhn. Mat. Nauk **1980**, no. 2, 27–31; English transl. in this volume.

2. V. P. Soltan and P. S. Soltan, *d-convex functions*, Dokl. Akad. Nauk SSSR **249** (1979), 555–558; English transl. in Soviet Math. Dokl. **20** (1979).

3. L. M. Blumenthal, *Theory and applications of distance geometry*, Clarendon Press, Oxford, 1953.

4. V. P. Soltan and P. S. Soltan, *A criterion for normability for a class of linear metric spaces*, Mat. Issled. **10** (1975), no. 2 (36), 275–279. (Russian)

5. R. Tyrrell Rockafellar, *Convex analysis*, Princeton Univ. Press, Princeton, N.J., 1970.

6. A. Wayne Roberts and Dale E. Varberg, *Convex functions*, Academic Press, 1973.

7. V. P. Soltan, *On metrically extremal and boundary points*, Modern Geometry. Questions of Differential Geometry (L. A. Berner, editor), Izdat. Leningrad. Gos. Univ., Leningrad, 1980, pp. 64–74. (Russian)

8. ——, *Metrically extremal and boundary points of convex sets in a normed space*, Mat. Issled. Vyp. 54 (1980), 155–163. (Russian)

9. L. F. German, V. P. Soltan, and P. S. Soltan, *Some properties of d-convex sets*, Dokl. Akad. Nauk SSSR **212** (1973), 1276–1279; English transl. in Soviet Math. Dokl. **14** (1973).

Translated by RALPH P. BOAS

Amer. Math. Soc. Transl.
(2) Vol. **134**, 1987

On a Class of Singular Two-Dimensional Integral Equations

UDC 517.968.2

B. M. BIL'MAN

Using the complex notation $z = re^{i\varphi}$, $\varsigma = \rho e^{i\theta}$ for the points in the plane, we define an operator Q by

$$\begin{aligned}(Qf)(z) &\equiv \frac{z}{\bar z} \iint\limits_{|\varsigma|\le 1} \frac{k(r,\rho) f(\varsigma)}{(\varsigma - z)^2}\, dS_\varsigma \\ &= \frac{z}{\bar z} \lim_{\varepsilon\to 0} \Bigg\{ \iint\limits_{|\varsigma|<|z|-\varepsilon} + \iint\limits_{|z|+\varepsilon<|\varsigma|\le 1} \Bigg\},\end{aligned} \tag{1}$$

where $k(r,\rho)$ is a bounded measurable function, $f(\varsigma)$ is an integrable function, $0 < |z| < 1$, and dS_ς is the area element. For $z = 0$ and $|z| = 1$ the left-hand side of (1) is defined by its limit values if they exist. All such two-dimensional integral operators encountered below are understood in the sense of (1).

Consider the equation

$$f(z) - (Qf)(z) = g(z), \qquad |z| \le 1. \tag{2}$$

The case $k(r,\rho) \equiv \lambda = \text{const}$ was studied in [1],([1]) where there are also comparisons with closely related publications by others. In this note we present results of an investigation of the operator Q and equation (2) in the complex Banach function spaces $\tilde C^0_\beta$, $\tilde C_\beta$, $\tilde M_\beta$, and $\tilde L^p_{\beta-2/p}$ described in [1]. We give definitions for them.

1980 *Mathematics Subject Classification* (1985 *Revision*). Primary 45E05; Secondary 30E20.

Translation of Dokl. Akad. Nauk Tadzhik. SSR **22** (1979), 515–520; MR **81i**:45012.

([1])Incidentally, Theorem 3 in [1] should contain "$\lambda \notin \Gamma_\beta(m)$" instead of "$\lambda \in \Gamma_\beta(m)$", part a) of Theorem 4 should contain "$\Gamma_\beta(2)$" instead of "$\Gamma_\beta(z)$", and the reference on p. 5 should be [2] instead of [1].

Setting

$$f_m(r) = \frac{1}{2\pi} \int_0^{2\pi} f(z) e^{-im\varphi}\, d\varphi, \qquad m = 0, \pm 1, \pm 2, \dots,$$

we say that:

$f(z) \in \tilde{C}_\sigma$ (σ a real number) if $F(z) = |z|^\sigma f(z)$ is continuous for $|z| \le 1$, and

$$\|f\|_{\tilde{C}_\sigma} = \sum_{m=-\infty}^{\infty} \max_{0 \le r \le 1} |r^\sigma f_m(r)| = \sum_{m=-\infty}^{\infty} \max_{0 \le r \le 1} |F_m(r)| < \infty;$$

$f(z) \in \tilde{C}_\sigma^0$ if $f(z) \in \tilde{C}_\sigma$ and $F(0) = 0$;

$f(z) \in \tilde{M}_\sigma$ if $F(z)$ is measurable and bounded almost everywhere, and

$$\|f\|_{\tilde{M}_\sigma} = \sum_{m=-\infty}^{\infty} \operatorname*{ess\,sup}_{0 \le r \le 1} |F_m(r)| < \infty;$$

$f(z) \in \tilde{L}_\sigma^p$ if $F(z)$ is measurable and has integrable pth power, $1 \le p < \infty$, and

$$\|f\|_{\tilde{L}_\sigma^p} = \sum_{m=-\infty}^{\infty} \left\{ \int_0^1 r|F_m(r)|^p\, dr \right\}^{1/p} < \infty.$$

As in [1], we introduce the common notation $\tilde{E}_\beta$ for the spaces $\tilde{C}_\beta^0$, $\tilde{C}_\beta$, $\tilde{M}_\beta$ and $\tilde{L}_{\beta-2/p}^p$ with $0 < \beta < 2$ and $1 \le p < \infty$.

THEOREM 1. *Let $k(r, \rho)$ be continuous for $0 \le r$, $\rho \le 1$.*

a) *The operator Q is bounded on $\tilde{E}_\beta$, with*

$$\|Q\|_{\tilde{E}_\beta} \le \begin{cases} \dfrac{2\pi \max |k(r,\rho)|}{\beta}, & 0 < \beta \le 1, \\[2ex] \dfrac{2\pi \max |k(r,\rho)|}{2-\beta}, & 1 \le \beta < 2. \end{cases}$$

b) *If $k(\rho, \rho) = 0$, $0 \le \rho \le 1$, then Q is a compact operator on $\tilde{E}_\beta$, and it maps $\tilde{M}_\beta$ into $\tilde{C}_\beta^0$.*

Assertion a) can be proved by straightforward estimates of the Fourier coefficients, with respect to the polar angle, of the value of the operator. The proof of b) is based on uniform approximation of $k(r, \rho)$ by functions satisfying a Hölder condition.

REMARK 1. For boundedness of the operator Q on $\tilde{M}_\beta$ and $\tilde{L}_{\beta-2/p}^p$ it suffices to require that $k(r, \rho)$ be measurable and bounded. We remark aslo that if $f \in \tilde{E}_\beta$, then the limits of each of the terms in the curly brackets in (1) exist separately (at least for almost all r with $0 < r < 1$), and the limits are uniform with respect to $\varphi = \arg z$.

Theorem 1 shows that the equation

$$f(z) - \frac{z}{\bar{z}} \iint_{|\varsigma| \le 1} \frac{k(\rho,\rho) f(\varsigma)}{(\varsigma - z)^2}\, dS_\varsigma = g(z), \qquad |z| \le 1, \tag{3}$$

can be taken as the characteristic equation for (2), and we proceed to a consideration of this equation.

The investigation of (3) (as in [1] and [2]) is based on a transition to one-dimensional integral equations for the Fourier coefficients of the unknown function with respect to the polar angle; these coefficients have the form

$$f_m(r) = \frac{2\pi(1-m)}{r} \int_0^r \left(\frac{\rho}{r}\right)^{1-m} k(\rho,\rho) f_m(\rho)\, d\rho + g_m(r), \qquad m \le 0,$$

$$f_m(r) = \frac{2\pi(m-1)}{r} \int_r^1 \left(\frac{\rho}{r}\right)^{1-m} k(\rho,\rho) f_m(\rho)\, d\rho + g_m(r), \qquad m \ge 1. \quad (4)$$

Let

$$\mu_\beta(m) = \frac{2-m-\beta}{2\pi(1-m)}, \qquad m = 0, -1, \pm 2, \dots, \quad 0 < \beta < 2,$$

$$\mu_\beta(1) = +\infty \quad \text{for } \beta \ne 1.$$

The results in [3] and [4] are applicable to each of the equations (4), and they give us that the mth equation ($m \ne 1$) is normally solvable if and only if $\operatorname{Re} k(0,0) \ne \mu_\beta(m)$. If $\operatorname{Re} k(0,0) < \mu_\beta(m)$, then the mth equation is unconditionally uniquely solvable. If $\operatorname{Re} k(0,0) > \mu_\beta(m)$, then for $m \le 0$ the homogeneous equation has exactly one linearly independent solution, which is used to construct a solution of the homogeneous equation (3) of the form $f_m(r)e^{im\varphi}$, and the nonhomogeneous equation is unconditionally solvable; for $m \ge 2$ the homogeneous equation does not have nontrivial solutions, and a single condition is necessary and sufficient for the nonhomogeneous equation to be solvable. Furthermore, the solutions are solvability conditions for the equations (4) can be written out explicitly. This is the starting information for our investigation of the character of solvability of (3) in $\tilde{E}_\beta$. First of all, it is not hard to establish that the conditions $\operatorname{Re} k(0,0) \ne 1/2\pi$ and $\operatorname{Re} k(0,0) \ne \mu_\beta(m)$, $m = 0, -1, \pm 2, \dots$, are necessary for normal solvability. Suppose that they hold. Then if it is assumed in addition that $\operatorname{Re} k(\rho,\rho) \ne 1/2\pi$, $0 < \rho \le 1$, it can be shown by using estimates of particular solutions of (4) in terms of the free terms that the Fourier series constructed from these solutions converges in the $\tilde{E}_\beta$-norm to the desired function $f(z)$, and equation (3) is normally solvable. The investigation of (3) is concluded by proving that the condition $\operatorname{Re} k(\rho,\rho) \ne 1/2\pi$, $0 < \rho \le 1$, is necessary for normal solvability. This proof is carried out in principle by constructing a sequence of free terms for which (3) is solvable, but the limit of the sequence does not belong to the range of the operator on the left-hand side of (3) if the indicated condition fails to hold. In some cases this construction is preceded by a suitable uniform approximation of the function $k(\rho,\rho)$.

THEOREM 2. *Suppose that $k(\rho,\rho)$ is continuous. Then equation* (3) *is normally solvable in $\tilde{E}_\beta$ if and only if*

$$\operatorname{Re} k(\rho,\rho) \ne \frac{1}{2\pi}, \quad 0 \le \rho \le 1; \qquad \operatorname{Re} k(0,0) \ne \mu_\beta(m), \quad m = 0, -1, \pm 2, \dots.$$

The passage from (3) to the general equation (2) is based on Theorem 1, with use of certain general facts in the theory of linear operators (see [5]). Furthermore, for these facts to be applicable in the cases when both characteristics of equation (3) (the number of linearly independent solutions of the homogeneous equation and the number of solvability conditions for the nonhomogeneous equation) are infinite, we represent $\tilde{E}_\beta$ as the direct sum of the subspaces $\tilde{E}_\beta^+$ (consisting of the functions $f(z)$ such that $f_m(r) \equiv 0$ for $m \leq 0$) and $\tilde{E}_\beta^-$ ($f_m(r) \equiv 0$ for $m \geq 1$), which are invariant under Q. In each of these subspaces one of the characteristics of (3) is always finite (more precisely, equal to zero). Moreover, information from [4] about the character of solvability of the equations corresponding to (2) for the Fourier coefficients of the desired function is also used. These equations are obtained from (4) by replacing $k(\rho, \rho)$ by $k(r, \rho)$.

THEOREM 3. *Let $k(r, \rho)$ be continuous. All the assertions of Theorem 2 remain in force for equation (2) except perhaps for the assertion that the condition* $\operatorname{Re} k(0,0) \neq \mu_\beta(m)$, $m = 0, -1, -2, \ldots$, *is necessary for normal solvability in $\tilde{C}_\beta^0$ or $\tilde{C}_\beta$ when $\beta \neq 1$.*

In the case of normal solvability denote by α_β the dimension of the subspace of solutions of the homogeneous equation (2), and by γ_β the number of necessary and sufficient conditions for solvability of the nonhomogeneous equation in $\tilde{E}_\beta$. Obviously, α_β and γ_β are at most countable.

THEOREM 4. *Suppose that $k(r, \rho)$ is continuous and that* $\operatorname{Re} k(\rho, \rho) \neq 1/2\pi$, $0 \leq r, \rho \leq 1$. *Then the following assertions are true*:

a) *If $0 < \beta < 1$, then for* $\operatorname{Re} k(0,0) < \beta/2\pi$ *equation* (2) *is unconditionally uniquely solvable in $\tilde{E}_\beta$, and if* $\mu_\beta(m_0) < \operatorname{Re} k(0,0) < \mu_\beta(m_0 + 1)$, *then $\alpha_\beta = 0$ and $\gamma_\beta = m_0 - 1$ when $m_0 = 2, 3, \ldots$, and α_β and γ_β are infinite when $m_0 = 0, -1, -2, \ldots$.*

b) *If $1 < \beta < 2$, then for* $\operatorname{Re} k(0,0) < (2 - \beta)/2\pi$ *equation* (2) *is unconditionally uniquely solvable in $\tilde{E}_\beta$, and if* $\mu_\beta(m_0 + 1) < \operatorname{Re} k(0,0) < \mu_\beta(m_0)$, *then $\alpha_\beta = |m_0|$ and $\gamma_\beta = 0$ when $m_0 = -1, -2, \ldots$, and α_β and γ_β are infinite when $m_0 = 1, 2, \ldots$.*

c) *If $\beta = 1$, then $\alpha_1 = \gamma_1 = 0$ when* $\operatorname{Re} k(0,0) < 1/2\pi$, *and α_1 and γ_1 are infinite when* $\operatorname{Re} k(0,0) > 1/2\pi$.

Moreover, if $\alpha_\beta > 0$, then there exists a basis of solutions of the homogeneous equation (2) *which is common for all the spaces in the series $\tilde{E}_\beta$, and for $\gamma_\beta > 0$ the solvability conditions for the nonhomogeneous equation have the form*

$$\iint\limits_{|z| \leq 1} g(z)\psi(z)\, dS_z = 0, \tag{5}$$

where $\psi(z)$ runs through a basis, common for the spaces in the series $\tilde{E}_{2-\beta}$, of solutions of the homogeneous equation that is the transpose of (2). *The elements of the first of these bases are the functions $f_m(r)e^{im\varphi}$ $(m \leq 0)$, and the elements of the second are the functions $\psi_n(r)e^{in\varphi}$ $(n \leq -2)$, where $f_m(r)$ and $\psi_m(r)$*

satisfy the corresponding one-dimensional homogeneous equations for the Fourier coefficients, each of which has at most one linearly independent solution.

REMARK 2. Instead of the continuity of $k(r, \rho)$ it suffices to require that it be bounded and measurable, be continuous for $|r-\rho| \leq \delta$ $(0<\delta<1)$, and satisfy in addition the condition

$$\lim_{r \to r_0} \int_0^1 |k(r,\rho) - k(r_0,\rho)|\, d\rho = 0, \qquad 0 < r_0 \leq 1,$$

for the spaces $\tilde{C}^0_\beta$ and $\tilde{C}_\beta$;

$$\lim_{\Delta\rho \to 0} \int_0^1 |k(r,\rho+\Delta\rho) - k(r,\rho)|\, d\rho = 0$$

uniformly with respect to $r \in [\eta, 1]$, $0 < \eta < 1$, for $\tilde{M}_\beta$, and

$$\lim_{\Delta r \to 0} \int_0^1 |k(r+\Delta r,\rho) - k(r,\rho)|\, dr = 0$$

uniformly with respect to $\rho \in [\eta, 1]$ for $\tilde{L}^1_{\beta-2}$.

It should be mentioned that the assertion about a common basis of solutions of the homogeneous equation (2) (of the homogeneous equation that is the transpose of (2)) applies now, of course, to the spaces $\tilde{E}_\beta$ (respectively, $\tilde{E}_{2-\beta}$) in which it is possible to consider the equation simultaneously for a given $k(r, \rho)$. Moreover, under the solvability conditions (5) the space $\tilde{L}^1_{-\beta}$ for the conjugate equation corresponds to the spaces $\tilde{C}^0_\beta$, $\tilde{C}_\beta$, and $\tilde{M}_\beta$ for the original equation, the space $\tilde{M}_{2-\beta}$ corresponds to $\tilde{L}^1_{\beta-2}$, and $\tilde{L}^p_{2-\beta-2/p}$ corresponds to $\tilde{L}^p_{\beta-2/p}$, $1 < p < \infty$.

REMARK 3. The scheme we have presented can also be used to investigate the equations with kernels

$$\frac{\varsigma}{\bar{\varsigma}} k(\rho, r)(\varsigma - z)^{-2}$$

(the transpose of (2)),

$$\frac{\bar{z}}{z} k(r,\rho)(\bar{\varsigma} - \bar{z})^{-2}, \ \frac{\varsigma}{\bar{z}} k(r,\rho)(\varsigma - z)^{-2}, \text{ etc.}$$

The author thanks Professor L. G. Mikhaĭlov for discussion and for useful advice.

BIBLIOGRAPHY

1. B. M. Bil'man and G. Dzhangibekov, *The solvability of a singular two-dimensional integral equation*, Dokl. Akad. Nauk Tadzhik. SSR **20** (1977), no. 4, 3–8. (Russian)

2. L. G. Mikhaĭlov, *On some two-dimensional integral equations with homogeneous kernels*, Dokl. Akad. Nauk SSSR **190** (1970), 272–275; English transl. in Soviet Math. Dokl. **11** (1970).

3. ———, *Integral equations with locally homogeneous kernels of degree* −1, Differential and Integral Equations with Singular Coefficients (L. G. Mikhaĭlov, editor), "Donish", Dushanbe, 1969, pp. 54–72.(Russian)

4. B. M. Bil′man, *Integral equations with variable limits of integration whose kernels have a singularity of the type of a homogeneous function of degree* -1, Differential and Integral Equations with Singular Coefficients (L. G. Mikhaĭlov, editor), "Donish", Dushanbe, 1969, pp. 19–40. (Russian)

5. I. Ts. Gokhberg [Israel Gohberg] and M. G. Kreĭn, *The basic propositions on defect numbers, root numbers, and indices of linear operators*, Uspekhi Mat. Nauk **12** (1957), no. 2(74), 43–118; English transl. in Amer. Math. Soc. Transl. (2) **13** (1960).

Translated by H. H. MCFADEN

Amer. Math. Soc. Transl.
(2) Vol. **134**, 1987

A Method of the Theory of Rings of Functions for the Construction of Compactifications

A. V. ZARELUA

In this communication we bring to light the connections between certain properties of a completely regular topological space X and the properties of the ring $C(X)$ of bounded continuous real-valued functions on X.

The basic general connections follow here from the Gel'fand-Shilov theory [1]: There exists a one-to-one correspondence between the closed subrings with unity of the ring $C(X)$ nd the compactifications of the continuous images of the space X. To every compactification of the space X there corresponds a closed subring K with unity which has the following property of regularity: If $x_1, x_2 \in X$, $x \neq x_2$, then there exists a function $f \in K$ such that $f(x_1) \neq f(x_2)$.

The following explicit construction of the space $\mathfrak{M}(K)$, corresponding in this theory to a closed subring $K \subseteq C(X)$ with unity, is often useful. If $K = \{f_\lambda\}_{\lambda\in\Lambda}$, then $\mathfrak{M}(K) = \overline{F_K(X)}$, where $F_K = \prod_{\lambda\in\Lambda} f_\lambda$ is the product of the functions $f_\lambda: X \to I_\lambda$, $I_\lambda = [\inf f_\lambda, \sup f_\lambda]$. Thus $\mathfrak{M}(X)$ can be considered as a subet of the Tychonoff cube $T = \prod_{\lambda\in\Lambda} I_\lambda$. The inverse correspondence is the following: to a compactification $\alpha[F(X)]$ of the continuous image $F(X)$ there corresponds the closed subring $K \subseteq C(X)$ with unity, determined by comparison with $\alpha[F(X)]$ of the subring K of the functions of the form $g \circ F$ on the space X with $g \in C(\alpha[F(X)])$. After this, it is natural to consider an arbitrary family of functions $M \subseteq C(X)$, the continuous mapping $F_M = \prod_{f\in M} f$ into the Tychonoff cube, and the compact set $B_M = \overline{F_M(X)}$. Using the Stone-Weierstrass theorem, it is easy to see that B_M corresponds to the smallest closed subring $[M]$ with unity containing the set M of the ring $C(X)$, i.e., to the subring $[M]$ obtained

1980 *Mathematics Subject Classification* (1985 *Revision*). Primary 54C40, 54D35; Secondary 54C20, 46B05.

Translation of Contributions to Extension Theory of Topological Structures (Proc. Sympos., Berlin, 1967; J. Flachsmeyer et al., editors), VEB Deutscher Verlag Wiss., Berlin, 1969, pp. 249–256; MR **40** #2006.

by taking the closure in $C(X)$ of the ring $R[M]$ of polynomials in the functions $f \in M$ with real coefficients. For example, if (X, Δ) is a proximity space and M is the family (ring) of all bounded Δ-functions, then B_M is the compactification of the space X corresponding, in the theory of Yu. M. Smirnov, to the proximity space (X, Δ) (S. V. Fomin [2], Yu. M. Smirnov [3]).

Incidentally, this construction allows us to reduce the proof of the Stone-Weierstrass theorem to the classical case, namely the Weierstrass theorem for functions of several variables. Indeed, it follows from this that it is sufficient to prove this theorem for the case of the Tychonoff cube $T = \prod_\lambda I_\lambda$ and that subring of $C(T)$ which is generated by the projections $T \to I_\lambda$. But it is easily seen from the description of the topology in T that the continuous functions on T can be uniformly approximated by functions whose values depend only on a finite number of coordinates. Hence the statement follows from the classical theorem of Weierstrass.

In problems concerning the relation between the weight of a space X and the weights of the compactifications of its continuous images, the following additional obervation is useful: If M is a family of functions such that $M = Q[M]$ ($Q[M]$ is the set of polynomials in the functions of the family M with rational coefficients), then M is everywhere dense in $C(B_M)$. Here M is considered, in a natural way, as a subset of the ring $C(B_M)$. This observation is an almost obvious consequence of the denseness of the set of rational numbers in R. Let us also obseve that the weight of a compact space X is equal to the weight of the ring $C(X)$ if by the latter is understood the least transfinite cardinality of a family E generating the whole ring, i.e. $[E] = C(X)$.

§1. The extension of mappings

Let us consider the following problem: Given a family $F = \{F_\lambda\}$ of mappings of a space X into compact spaces Y_λ, construct a compactification αX to which all these mappings can be extended. Clearly, there exists at least one compactification with this property: for example, the Stone-Čech compactification βX. It is also clear that, in general, the weight of αX must be approximately equal to

$$\tau_F = \max\{\text{Weight of } X,\ \text{cardinality of } F \times \max_\lambda\{\text{weight of } Y_\lambda\}\}.$$

This problem of the extension of mappings to the compactification $\alpha_F X$ of weight $\leq \tau_F$ is easily solved in the following way: $\alpha_F X = \overline{\Psi X}$, where $\Psi\colon X \to X \times \prod_\lambda Y_\lambda$ is the product $\Psi = 1_X \times \prod F_\lambda$ of mappings. The mappings extending F_λ will then simply be the projections of the set $\overline{\Psi(X)}$ onto the factors Y_λ. From the point of view of the ring $C(X)$ this construction corresponds to the consideration of the closed subring with unity of $C(X)$ generated by a regular system of functions whose cardinality is equal to the weight of the space X and every system E_λ of functions of minimum cardinality among all the systems of generators of a subring isomorphic to the images of the ring $C(Y_\lambda)$ under the counter homomorphisms $F_\lambda^*\colon C(Y_\lambda) \to C(X)$.

A similar problem has been considered by Engelking [4]: Given a family F of mappings $F_\lambda: X \to X$, construct a compactification αX such that every mapping F_λ can be extended to a mapping $\tilde{F}_\lambda: \alpha X \to \alpha X$. For its solution it is sufficient to take the regular family of functions $\{\pi_\alpha\}$ and to consider the family Φ of functions of the form $\pi_\alpha \circ F_{\lambda_1} \circ F_{\lambda_2} \circ \cdots \circ F_{\lambda_k}$ and the compact space B_Φ; $\tilde{F}_\lambda$ is the mapping corresponding to that mapping of the family Φ into itself which is defined by simple composition:

$$\pi_\alpha \circ F_{\lambda_1} \circ \cdots \circ F_{\lambda_k} \to \pi_\alpha \circ F_{\lambda_1} \circ \cdots \circ F_{\lambda_k} \to \pi_\alpha \circ F_{\lambda_1} \circ \cdots \circ F_{\lambda_k} \circ F_\lambda.$$

§2. The extension of mappings to the compactifications with special dimensional properties and the factorization theorems

We know the following property of the dimension of a normal space X due to P. S. Aleksandrov: $\dim X \leq n$ if and only if every continuous mapping of a closed subset $F \subseteq X$ into the sphere S^n can be extended to the whole space X. On the other hand, for a finite or countable group G and a paracompact space X the results of Huber [5] and Borsuk's lemma show that if the cohomological dimension $\operatorname{cdim}_G X$ is defined as the smallest number n such that $\check{H}^i(X, F; G) = 0$ for every $i > n$ and every closed subset $F \subseteq X$ (here $\check{H}^i(X, F; G)$ denotes the spectral group of the Aleksandrov-Čech cohomology of the pair (X, F)), then $\operatorname{cdim}_G X \leq n$ if and only if every mapping of each closed subset $F \subseteq X$ into the Eilenberg-MacLane complex $K(G, n)$ can be extended to the whole of X. Let us define $\dim_G X$ of a normal space X as $\operatorname{cdim}_G \beta X$; $\dim_G X$ can also be given an internal characterization with the help of the cohomology of the nerves that $\dim_G X = \operatorname{cdim}_G X$ if X is compact, or if X is paracompact, $\dim X < \infty$ and the group is finitely generated; the latter follows from the fact that in the case of a finitely generated group G every finite-dimensional skeleton of the complex $K(G, n)$ is a finite polyhedron.

GENERAL THEOREM ON COMPACTIFICATIONS. *Let there be given a family $\{X_\lambda\}$ of cardinlity at most τ of closed subsets of a normal space X, a family $\{F_\alpha\}$ of cardinality at most τ of mappings of X into the compact spaces Y_α of weights at most τ, a family of transformations $T_\omega: X \to X$ of cardinality at most τ, and the numbers $n_{\lambda,\sigma}$, the dimensions of the set X_λ relative to countable or finite groups G_σ; the cardinality of the family $\{G_\sigma\}$ is also assumed at most τ. Then each compactification bX of the space X is followed by a compactification aX of the space X of weight at most* max{*Weight of bX, τ*} *such that the following conditions hold:*

1. *The mappings $F_\alpha: X \to Y_\alpha$ can be extended to mappings $\tilde{F}_\alpha: aX \to Y_\alpha$.*
2. *The transformations $T_\omega: X \to X$ can be extended to transformations $\tilde{T}_\omega: aX \to aX$.*
3. *The dimensions of the closure* $\dim_{G_\sigma} \operatorname{Cl}_{aX} X_\lambda = n_{\lambda,\sigma}$.
4. $\dim \operatorname{Cl}_{aX} X_\lambda = \dim X_\lambda$.

GENERAL FACTORIZATION THEOREM. *Under the assumption of the general theorem on compactifications there exist a compact space B of weight at most τ, a mapping $F: X \to B$, mappings $G_\alpha: B \to Y_\alpha$, and transformations $U_\omega: B \to B$ such that*

1. $G_\alpha \circ F = F_\alpha$,
2. $U_\omega \circ F = F \circ T_\omega$,
3. $\dim_{G_\sigma} \mathrm{Cl}_B FX_\lambda = n_{\lambda,\sigma}$, *and*
4. $\dim \mathrm{Cl}_B FX_\lambda = \dim X_\lambda$.

Let us make some remarks and deduce some corollaries of these theorems.

1. Similar results in this direction were obtained independently and earlier by I. A. Shvedov.

2. The second theorem is stronger than the first: If the embedding mapping $X \to bX$ is also adjoined to the system $\{F_\alpha\}$ in the second theorem, then the compact space B is the required aX.

3. These theorems, excluding the questions related to points 2) and 3), were proved by the author in [6].

4. For the space X with a countable base and a countable family $\{X_i\}$ of its closed subsets there exists a compactification aX such that $\dim \mathrm{Cl}_{aX} X_i = \dim X_i$. This is the classical theorem of Hurewicz.

5. Every normal space with a distinguished countable family $\{X_i\}$ of its closed subsets has a compactification aX of the same weight as X such that $\dim \mathrm{Cl}_{aX} X_i = \dim X_i$ (Sklyarenko [7]).

6. If in the theorem on compactifications we take $\{X_\lambda\} = \{X\}$ and the empty families $\{F_\alpha\}$ and $\{G_\sigma\}$, then we obtain the results of Engelking and Sklyarenko [8].

7. If in the factorization theorem we take $\{X_\lambda\} = \{X\}$ and $\{Y_\alpha\} = \{Y\}$ and assume the families $\{T_\omega\}$ and $\{G_\sigma\}$ are empty, then we obtain the theorem of Mardešić [9].

8. Forge's theorem [10] is also a corollary to the theorem on compactifications.

We given here the proof of the second theorem as illustrating what we have said above. Let us put $K_{\sigma,\lambda} = K(G_\sigma, N_{\lambda,\sigma})$ and let $P^s_{\sigma,\lambda}$ be the family (obviously countable) of all finite polyhedra which are the sums of a finite number of cells of the CW-complex $K_{\sigma,\lambda}$. The polyhedra $P^s_{\sigma,\lambda}$ will be regarded as lying in the Euclidean space $E^{N(s,\sigma,\lambda)}$ of an appropriate dimension. Let us take a neighborhood $U^s_{\sigma,\lambda}$ of the polyhedron $P^s_{\sigma,\lambda}$ in this Euclidean space, for which there exists a deformation retraction $\mathrm{Ret}^s_{\sigma,\lambda}: \overline{U}^s_{\sigma,\lambda} \to P^s_{\sigma,\lambda}$. Analogously, let $\mathrm{Ret}_\lambda: \overline{U}_\lambda \to S^{\dim X_\lambda}$ be a retraction of the closure of some neighborhood $\overline{U}_\lambda$ of the space $S^{\dim X_\lambda}$. Let us choose a regular family of functions on Y_α of minimum cardinality; we denote its image in $C(X)$ under the mapping $F^*_\alpha: C(Y_\alpha) \to C(X)$ by M_α. We shall denote the set of all finite compositions of the form $a \circ T_{\omega_1} \circ \cdots \circ T_{\omega_k}$, $a \in A \subset C(X)$, by $T(A)$, where $T = \{T_\omega\}$. Let us put $M^1 = Q[T(\bigcup_\alpha M_\alpha)]$. By induction we shall construct families M^j of functions with the following properties.

A) $M^{j'} \subseteq M^{j''}$ if $j' \leq j''$.

B) $Q[M^j] = M^j$.

C) $T(M^j) = M^j$.

D) For every mapping of the form $\mathrm{Ret}^s_{\sigma,\lambda} \circ \Omega/\Phi$, where the components of the mapping Ω belong to M^{j-1}, of a closed set $\Phi \subseteq X_\lambda$ into one of the polyhedra $P^s_{\sigma,\lambda}$ there exists a mapping of the whole of X into some $R^{N(s',\sigma,\lambda)}$, whose restriction to X_λ is a mapping of X_1 into $K^{s'}_{\sigma,\lambda}$ and whose components belong to M^j.

E) For every mapping of the form $\mathrm{Ret}_\lambda \circ \Omega/\Phi$, where the components of Ω belong to M^{j-1}, of a closed set $\Phi \subseteq X_\lambda$ into the sphere $S^{\dim X_\lambda}$ there exists a mapping of the whole of X into $E^{\dim X_\lambda + 1}$ extending $\mathrm{Ret}_\lambda \circ \Omega$ to X_λ whose components belong to M^j.

F) The cardinality of M^j is at most τ.

If $M^1 \subseteq \cdots \subseteq M^i$ have already been constructed, then we adjoin to M^i the set N of functions from whose elements we can compose mappings into Euclidean spaces satisfying the conditions D) and E). We can always find such functions for D) on X_λ, since by assumption

$$\dim_{G_\sigma} X_\lambda = \dim_{G_\sigma} \beta X_\lambda = n_{\lambda,\sigma}$$

(see the characterization of the dimension by the possibility of extension of mappings), and then it is sufficient to extend these functions to the whole of X. Here we use the well-known fact that every compact subset of the CW-complex $K_{\sigma,\lambda}$ intersects only a finite number of cells, and, in particular, the image of βX_λ under an extended mapping lies in a $P^s_{\sigma,\lambda}$. The functions for E) are found analogously using Aleksandrov's criterion. After this we assume $M^{i+1} = Q[T(M^i \cup N)]$. Let us put $M = \bigcup_1^\infty M^j$. Then M has all the properties B)–F) if we take M in place of M^j and M^{j-1}. Moreover, $M \supseteq M^1$. We shall show that $B = B_M$ has almost all the required properties. First of all, let us choose in $C(B)$ a set M' isomorphic to M. If we put $F = F_M$, then a correspondence between $f' \in M'$ and $f \in M$ given by the equality $f = f' \circ F$. Let $G_\alpha = \prod_{f' \in M'_\alpha} f'$, and let the transformation $U_\omega\colon B \to B$ be determined by the transformation $U'_\omega\colon M' \to M'$, $U'_\omega(m) = m \circ T_\omega$. Then properties 1) and 2) of the theorem and the inequality Weight of $B \leq \tau$ are obvious. Now let Φ be a closed subset of the compact space $\mathrm{Cl}_B\, FX_\lambda$ and Ω a mapping of Φ into $K_{\sigma,\lambda}$; Ω can be regarded as a mapping into one of the polyhedra $P^s_{\sigma,\lambda}$. Since $M' = Q[M']$ is dense in $C(B)$, there exists a mapping Ω' of some closed neighborhood $\overline{V}$ of the set Φ into $U^s_{\sigma,\lambda}$ whose components belong to M, such that $\mathrm{Ret}^s_{\sigma,\lambda} \circ \Omega'$ is homotopic to Ω. According to D) there exists an extension ξ to X_λ of the mapping

$$\mathrm{Ret}^s_{\sigma,\lambda} \circ \Omega' | FX_\lambda$$

(or, in terms of $C(X)$, an extension of the mapping $\mathrm{Ret}^s_{\sigma,\lambda} \circ \Omega' \circ F | F^{1}(\overline{V} \cap FX))$ into another polyhedron $P^{s'}_{\sigma,\lambda} \subseteq K_{\sigma,\lambda}$ whose components belong to M'. In particular, this mapping ξ is automatically extended to $\mathrm{Cl}_B\, FX_\lambda$. Since $\xi|\Phi$ is equal to $\mathrm{Ret}^s_{\sigma,\lambda} \circ \Omega'|\Phi$, which is homotopic to Ω, according to Borsuk's lemma there exists an extension of Ω also. Thus it has been proved that $\dim_{G_\sigma} \mathrm{Cl}_B FX_\lambda \leq n_{\lambda,\sigma}$. Analogously, $\dim \mathrm{Cl}_B FX_\lambda \leq \dim X_\lambda$. Finally, the equality would have been easily obtained if from the very beginning we had included in the family $\bigcup_\alpha M_\alpha$ all those mappings whose restrictions to some closed subsets $\Phi_\lambda \subseteq X_\lambda$ give mappings into $K(G_\sigma, n_{\lambda,\sigma} - 1)$ and into $S^{\dim X_\lambda - 1}$, which cannot be extended to the whole of X_λ. The theorem is proved.

3. Comparison with other constructions of compactifications

A. If μ is a uniform structure on a space X having a base consisting of finite open converings, then the compactification of X, i.e. the completion relative to this uniform structure, is, from the point of view of the ring $C(X)$, the extension determined by the ring of functions on X which are uniformly continuous relative to μ.

B. If the compactification αX is obtained as the limit of the inverse spectrum of the compacta B_λ, then, from the point of view of the ring $C(X)$, to this spectrum there corresponds the set, partially ordered by inclusion, of closed subrings $C_\lambda = \pi_\lambda^* C(\overline{\pi_\lambda X})$, where $\pi_\lambda \colon X \to B_\lambda$ is the canonical projection, and for arbitrary λ and μ there exists a $\nu \geq \lambda$, μ for which $C_\lambda \cup C_\mu \subseteq C_\nu$. The ring $C(\alpha X) \subseteq C(X)$ is simply the closure $\overline{\bigcup_\lambda C_\lambda}$ This observation permits us, if we wish, to substantially simplify the proofs of many theorems based on the technique of spectra.

Let us also observe tht the structure (in the sense of the theory of partially ordered sets) of all compactifications is isomorphic to the structure of the set of all closed regular subrings with unity of the ring $C(X)$:

$$C(\alpha X \vee \alpha' X) = C(\alpha X) \cap C(\alpha' X)$$

and $C(\alpha X \wedge \alpha' X)$ is the smallest closed subring containing $C(\alpha X) \cup C(\alpha' X)$.

§4. Spaces of real ideals

For an arbitrary real normed ring C with unity, following [11] we call every ideal I, the quotient modulo which is isomorphic to the normed field of the real numbers R, a *real ideal* (*h-ideal* in the terminology of [12]). Note that we do not require the completeness of the ring G. Every real ideal is maximal; the converse is not true. We introduce the intrinsic topology in the set $\mathfrak{M}(C)$ of all real ideals so that $\mathfrak{M}(C)$ becomes a compact space [12]. For example, for the free normed ring G on τ symbols, representing the ring of polynomials whose norm is defined as the sum of the moduli of the coefficients, the space $\mathfrak{M}(C)$ is a Tychonoff cube of weight τ. The basic facts about the spaces of real ideals stated in [12] are similar to analogous statements for the Gel′fand-Shilov maximal ideal spaces [1]. For example, to the homomorphism $C \to C'$ there corresponds the mapping $\mathfrak{M}(C') \to \mathfrak{M}(C)$. This mapping is an imbedding if the image of the ring C is everywhere dense in C'. In particular, the fact that a free ring can be homomorphically mapped onto any ring means that a compact space of weight at most τ can be imbedded in a Tychonoff cube of weight τ.

The use of the space of real ideals often allows us to use algebraic and functional-analytical methods for the construction of compacta with the required properties. For this it is necessary to have a description of these properties in sufficiently simple terms.

Following Katětov, we say that a subring $A \subseteq C$ is *algebraically closed in* C from $x^n + a_1x^{n-1} + \cdots + a_n = 0$, where $a_i \in A$ and $x \in C$, it follows that $x \in A$. It turns out that if the ring C is isomorphic to the subring $C(\mathfrak{M}(C))$, then A is algebraically closed in C if and only if the corresponding mapping $\mathfrak{M}(C) \to \mathfrak{M}(A)$ is zero-dimensional in the sense that $\dim f^{-1}(y) \leq 0$ for every y. These results can be useful for finding new approaches to the well-known problems of Hopf and Wilder about open zero-dimensional mappings of the n-dimensional cube Q^n, since $Q^n = \mathfrak{M}(C_n)$, where C_n is the free normed ring on n symbols.

We shall give here only two applications. These are theorems proved initially by B. A. Pasynkov.

A. Let X be a compact space, and $\pi = \{\pi_\lambda\}$ a regular system of functions of cardinality equal to the weight of X. For every $\tau \geq$ the weight of X and for every function $\varphi(\pi) \in Q[\pi_\lambda]$ let us take τ symbols $\tilde{\varphi}_\omega$. We shall consider the free normed ring $\tilde{C}$ on the symbols $\{\tilde{\pi}_\lambda\}$ and $\{\tilde{\varphi}_\omega\}$ assuming $\|\tilde{\pi}_\lambda\| = \|\pi_\lambda\|$ and $\|\tilde{\varphi}_\omega\| = \|\varphi(\pi)\|$. If I is the closed ideal generated by the elements of the form $\tilde{\varphi}_\omega^2 - \tilde{\varphi}_\omega\varphi(\tilde{\pi})$ and $C = \tilde{C}/I$, then $\mathfrak{M}(C)$ is the universal space in the class of all compacta of weight at most τ admitting a zero-dimensional mapping into the compact space X.

B. Let C' be the free normed ring on the symbols f, π_λ, and φ_λ; the cardinality of the set of indices is equal to τ. Let I be the closed ideal generated by the elements of the forms $\varphi_\lambda^2 - \varphi_n\pi_\lambda$ and $\pi_\lambda^2 - \pi_\lambda f$, and $C = C'/I$. Then $\mathfrak{M}(C)$ is a one-dimensional compact space admitting a zero-dimensional and open mapping onto a Tychonoff cube of weight τ and $T = \mathfrak{M}(\overline{C})$, where $\overline{C}$ is the free normed ring generated by the symbols $\{\pi_\lambda\}$. In particular, we obtain the following remarkable theorem of B. A. Pasynkov: *Every completely regular space is the clopen image of a one-dimensional space under a perfect mapping.*

BIBLIOGRAPHY

1. I. M. Gel′fand, D. A. Raĭkov and G. E. Shilov, *Commutative normed rings*, Fizmatgiz, Moscow, 1960; English transl., Chelsea, New York, 1964.

2. S. V. Fomin, *On the connection between proximity spaces and compactifications of completely regular spaces*, Dokl. Akad. Nauk SSSR **121** (1958), 236–238. (Russian)

3. Yu. M. Smirnov, *Generalization of the Stone-Weierstrass theorem to proximity spaces*, Czechoslovak Math. J. **10(85)** (1960), 493–500. (Russian; English summary)

4. R. Engelking, *Sur la compactification des espaces métriques*, Fund. Math. **48** (1959/60), 321–324.

5. Peter J. Huber, *Homotopical cohomology and Čech cohomology*, Math. Ann. **144** (1961), 73–76.

6. A. V. Zarelua, *The continuation of mappings to extensions with certain special properties*, Sibirsk. Mat. Zh. **5** (1964), 532–548. (Russian)

7. E. Sklyarenko, *On the imbedding of normal spaces into compact Hausdorff spaces of the same weight and dimension*, Dokl. Akad. Nauk SSSR **123** (1958), 36–39. (Russian)

8. R. Engelking and E. G. Sklyarenko, *On compactifications allowing extensions of mappings*, Fund. Math. **53** (1963), 65–79.

9. Sibe Mardešić, *On covering dimension and inverse limits of compact spaces*, Illinois J. Math. **4** (1960), 278–291.

10. A. B. Forge, *Dimension preserving compactifications allowing extensions of continuous functions*, Duke Math. J. **28** (1961), 625–627.

11. Leonard Gillman and Meyer Jerison, *Rings of continuous functions*, Van Nostrand, Princeton, N.J., 1960.

12. A. V. Zarelua, *A universal compact Hausdorff space with given weight and dimension*, Dokl. Akad. Nauk SSSR **154** (1984), 1015–1018; English transl. in Soviet Math. Dokl. **5** (1964).

Translated by Y. K. CHOUDHARY

Amer. Math. Soc. Transl.
(2) Vol. **134**, 1987

Extensions of Topological Structures and Metric Properties of Mappings

G. D. SUVOROV

We present here a brief survey of the main results obtained by the author and his students in the study of the metric properties of planar and spatial mappings of closed domains of Euclidean space and some other spaces. In our investigations the closure of a domain is obtained by adjoining boundary elements defined differently in different situations. A method basically originating in Carathéodory's well-known paper [1] on the prime ends of a simply connected domain, supplemented with a suitable metrization of closed domains with the help of "relative" distances, is used. The Carathéodory boundary was apparently first metrized by Mazurkiewicz [2], and the first estimate of the distortion of "relative" distances under conformal mappings of a closed disk onto a closed simply connected domain is due to Lavrent′ev [3]. In the work described here these ideas are systematically used and generalized in some respects. We investigate classes of mappings whose selection and study are, in our opinion, the next step after the study of the quasiconformal mappings, planar as well as spatial. The significance of the results is in the new facts of the theory of functions. From the topological point of view the work contains, perhaps, concrete realizations of relatively simple extensions of topological structures that are convenient in applications. However, this work again shows that the general topological notion of a boundary, going back to Carathéodory and systematically studied and extended to general topological structures by Myshkis [4], [5], finds an application also in the modern metric theory of mappings, and in the course of its further development the involvement of ever more general topological concepts in the

1980 *Mathematics Subject Classification* (1985 *Revision*). Primary 30-02, 30D40, 30G15, 54-02, 54C20; Secondary 30C60, 54E35.

Translation of Contributions to Extension Theory of Topological Structures (Proc. Sympos., Berlin, 1967; J. Flachsmeyer et al., editors), VEB Deutscher Verlag Wiss., Berlin, 1969, pp. 257–273.

use of the metric theory is inevitable. This gives us some motivation for the inclusion of our survey in the proceedings of a topological symposium.

Without aiming for full statements of all the results, we restrict ourselves to a general description of the topics and the exposition of the main directions in which the investigation is being carried out. The bibliography is not complete.

§1. General formulation of the problem

We are interested mainly in the metric properties of mappings of "closed" domains, where initially the mappings are given only in the interior of domains. Consequently, the first problem is about the possibility of extending a mapping to the boundary of the domain, in particular the problem of boundary correspondence if the mapping is a homeomorphism. It has two aspects: topological, the problem of the introduction of boundary elements in the domain, and metric, whether in fact for a given mapping a boundary correspondence is realized relative to the boundary elements introduced a priori. In this connection it is natural to start by restricting ourselves to the problem of selecting those classes of mappings for which we can introduce, in a uniform manner, boundary elements in admissible pairs of domains (i.e. in domains, each of which can be mapped onto the other by a mapping of the given class) such that the boundary correspondence within the class is realized precisely relative to these elements. Thus, it is necessary to have a compatible selection of the object: the class of mappings and the boundary elements in the admissible domains. Such "compatibility" is ensured if for a concrete class of mappings the following construction is successfully realized.

Let A be a domain in an arbitrary compactum (M, r) (M is a set of elements and r is a metric), B a domain in another compactum (N, ρ), and $T(A) = B$ a topological mapping. Let us introduce in A a new (relative) metric r_A and in B a metric ρ_B such that the relative metrics in A and B are respectively equivalent (in the small) to the metrics r and ρ of the above spaces. Let $\tilde{A}$ be the completion of A relative to r_A and $\tilde{B}$ the completion of B relative to ρ_B. The equivalence classes of Cauchy sequences in the spaces (A, r_A) and (B, ρ_B) are the "boundary" elements adjoined to A and B.

If we are given functions $\varphi_j(\alpha)$ $(j = 1, 2)$, $\varphi_j(\alpha) \to 0$ as $\alpha \to 0$, such that for any points $x', x'' \in A$ which are sufficiently close relative to r_A we have

$$\varphi_1[r_A(x', x'')] \le \rho_B[T(x'), T(x'')] \le \varphi_2[r_A(x', x'')], \tag{1}$$

then T can obviously be extended to a homeomorphism $T(\tilde{A}) = \tilde{B}$ (the problem of boundary correspondence is solved) and (1) is also valid in the "closed" domain:

$$\varphi_1[r_A(e_1, e_2)] \le \rho_B[T(e_1), T(e_2)] \le \varphi_2[r_A(e_1, e_2)], \tag{2}$$

where the e_j are elements of $\tilde{A}$.

The two-sided estimate (2), especially if φ_j and the "interval" on which (2) holds are the same for all the mappings of a certain class, contains information sufficient for finding the main metric properties of the mappings (the discovery of this information in the general case is the goal of an article by V. P. Luferenko and G. D. Suvorov, currently being prepared for publication).* Therefore, the search for "compatible" objects for which (2) (or at least a one-sided estimate) holds is the main problem of the investigations described in the sequel.

§2. Classes of mappings

Let E^n be n-dimensional Euclidean space with Cartesian coordinates $x = (x_1, \dots, x_n)$, and let $y \equiv T(x) = (f_1(x), \dots, f_p(x))$ be a mapping of a domain $D \subset E^n$ into E^p; let $\mu = \mu(y)$, $y = (y_1, \dots, y_p)$, be a continuous nonnegative function defined for all $y \in \Delta$, where Δ is a submanifold of E^p which contains the images of the domains D under the mappings T of a class to be defined later on.

Let $\alpha \geq 1$. For $\alpha > 1$ we assume that

$$\int_{\overline{\Delta}_1} \mu^{-1/(1-\alpha)}(y)\, dy < +\infty$$

for any bounded closed subdomain $\overline{\Delta}_1 \subset \Delta$, and for $\alpha = 1$ we assume that $\mu(y) \neq 0$ in Δ.

The mapping T belongs to the class $BL_k^\alpha(\mu)$ in D relative to the system of coordinates x, if the f_j ($j = 1, \dots, p$) are continuous in D and have generalized Sobolev partial derivatives of first order with respect to every x_i and, moreover, for any bounded closed subdomain $\overline{D}_1 \subset D$ we have

$$\int_{\overline{D}_1} \lambda(x, T)\, dx < \infty,$$

where

$$\lambda(x, T) = \sum_{i=1}^{n} \sum_{j=1}^{p} \left(\frac{\partial f_i}{\partial x_i} \right), \qquad dx = dx_1 \cdots dx_n,$$

and, moreover,

$$\int_D \mu[T(x)] \lambda^\alpha(x, T)\, dx \leq k.$$

The mapping $T \in BL^\alpha(\mu)$ in D if $T \in BL_k^\alpha(\mu)$ for some k. For $\mu \equiv 1$ and $\alpha = n/2$ the corresponding classes are denoted by $BL_k^{n/2}$ and $BL^{n/2}$. If $\mu \equiv 1$, $\alpha = 1$, and $n = p = 2$, we obtain the classes BL_k and BL, and if $\mu = 1/(1 + f_1^2 + f_2^2)^2$, $\alpha = 1$, and $n = p = 2$, we obtain the classes $\widetilde{BL}_k$ and $\widetilde{BL}$.

It is essential that the class $BL_k^\alpha(\mu)$ of mappings of the domain D is closed relative to uniform convergence in the interior of D (Baklanov and Suvorov [6]).

§3. Basic inequalities of the method

The method based on the use of the inequalities of the type of "the length-area principle", well known in the theory of analytic functions, is still the only

*Editor's note. In retrospect it appears that the author is referring to [36].

method for the study of the metric properties of the mappings of some of the classes defined above.

Let us consider a continuous mapping $w = T(z)$, $w = f_1 + if_2$, $z = x + iy$, of a domain $D \subset E^2$ into E^2. In the z-plane we consider the spherical metric (R) obtained by stereographic projection of the z-plane onto the Riemann sphere of radius R touching the plane at the origin. Analogously, in the w-plane we consider the spherical metric (r) (in general, $R \neq r$).

Let $F \subset E^2$ be a closed set, $\rho_R(z)$ the spherical distance from $z \in E^2$ to F, $\Gamma_t = \{z \colon \rho_R(z) = t\} \cap D$ the part of the set of spherical "level t" of the function ρ_R belonging to D, $q \geq 1$, and let f_j $(j = 1, 2)$ have generalized Sobolev derivatives of first order in D and be differentiable almost everywhere in D (the latter condiiton can be replaced by the condition of openness of the mapping T). Then

$$\int_{\tau_1}^{\tau_2} \frac{L_r^q[T(\Gamma_t)]}{L_R^{q-1}(\Gamma_t)}\, dt \leq \iint\limits_{D_{\tau_1\tau_2}} \mu_r^{q/2}[T(z)]\lambda^{q/2}(z, T)\mu_R^{(2-q)/2}(z)\, dx\, dy, \tag{3}$$

where τ_1 and τ_2 are numbers chosen in such a way that for $t \in [\tau_1, \tau_2]$ the set Γ_t is nonempty, $L_R(\Gamma_t)$ is the Hausdorff length determined on the basis of the spherical metric, $L_r[T(\Gamma_t)]$, for almost every t, is the sum of the spherical lengths (in the metric (r)) of the images of the constituent arcs of the set Γ_t, $D_{\tau_1\tau_2}$ is the subset of D covered by Γ_t as t varies and

$$\mu r(T) = \frac{1}{(1 + |T/2r|^2)^2}, \qquad \mu_R(z) = \frac{1}{(1 + |z/2R|^2)^2}.$$

However, in (3) in place of the Hausdorff length $L_R(\Gamma_t)$ we can take simply the spherical length in the metric (R) (see Ionin and Suvorov [7] and Kufarev and Nikulina [8]).

If F is a point, $q = 2$, and $T \in \widetilde{BL}$ in D, then from (3) we get

$$\int_{\tau_1}^{\tau_2} L_r^2[T(\Gamma_t)]\frac{dt}{t} \leq 2\pi \iint\limits_{D_{\tau_1\tau_2}} \mu[T(z)]\lambda(z, T)\, dx\, dy \tag{4}$$

which is the inequality in the class BL that has had till now the greatest number of applications (Lelong-Ferrand [9] and Suvorov [10], [11]). The more general result (3) has been obtained by Kufarev [12], [22].

There is a refinement of inequality (4) for Q-quasiconformal homeomorphisms and, in a somewhat modified form, for interior Q-quasiconformal mappings [10]. For interior Q-quasiconformal mappings an analogue of inequality (3) is given in [12].

An inequality which is an analogue of (4) for the three-dimensional case $(n = p = 3)$ has been estabished by Ovchinnikov and Suvorov in [13], and for an arbitrary $n \geq 3$ by Ovchinnikov in [14] as a corollary of a general metric-geometric inequality, estabished by him and having an "isoprometric" character.

We shall give the most general result ($n \geq 3$) of [14]:

Let $T \in BL^{n/2}$ in D; let $\{S_t\}$ be the family of concentric spheres in E^n centered at the point a, $a < \tau_1 \leq t \leq \tau_2$, and for this family let the sets $S'_t = S_t \cap D$ be nonempty. Denote by K_t the open spherical disc of spherical radius $R(t) \leq \pi t/2$, $K_t \subset S'_t$. Assume that on $[\tau_1, \tau_2]$ there are defined measurable functions $\Omega(t)$ and $\beta(t)$ such that for almost all t

$$\Omega(t) \leq \omega(K_t, T), \qquad \beta(t) \geq 2R(t)/\pi t,$$

where ω is the oscillation of T on K_t. Then

$$\int_{\tau_1}^{\tau_2} \frac{\Omega^n(t)}{t\beta(t)}\, dt \leq M_n \int_{D_{\tau_1\tau_2}} \lambda^{n/2}(x, T)\, dx, \tag{5}$$

where

$$D_{\tau_1,\tau_2} = \bigcup_{t\in[\tau_1,\tau_2]} S'_t, \qquad M_n = \frac{1}{2}\pi^n \Gamma\left(\frac{n-1}{2}\right) \left[\frac{\Gamma(1/2(n-1))}{\Gamma(n/2(n-1))}\right]^{n-1},$$

Γ being Euler's gamma function.

§4. Results in the case $n = p = 2$

The construction whose general scheme is given in §1 has been completely realized for this case [10]. As we have already observed, the beginning here was made by Lavrent′ev [3].

Inequality (4) serves for obtaining the basic inequalities (1). The compacta (M, r) and (N, ρ) are planes, completed by infinity; an admissible domain is an arbitrary simply connected plane domain with a boundary containing more than one point; $\{T\}$ is the class $\widetilde{BL}$ (or BL), where the inverse mapping must belong to the same class. r_A and ρ_B are the relative metrics introduced on the basis of the spherical or Euclidean metric. In the first case the definition is as follows: Let the simply connected domain $A \subset E^2$ contain the point O, and let $x', x'' \in A$, $x', x'' \neq 0$, be points. Then

$$\rho_A(x', x'') = \min(\rho_1, \rho_2),$$

where ρ_1 is the infimum of the spherical diameters of arcs joining x' and x'' in A, and ρ_2 is the infimum of the spherical diameters of cross cuts of A separating $x' \cup x''$ from O. The completion of A relative to ρ_A adjoins the point O and the "boundary" elements which can be identified with the Carathéodory prime ends. The functions φ_j in (1) are defined as $\varphi_2(\alpha) = K(\ln 1/\alpha)^{-1/2}$ and $\varphi_1 \equiv \widetilde{\varphi_2}^{-1}$ (inverse to the function φ_2), where K is a constant in the case of a class $\widetilde{BL}_k$ (or BL_k), i.e. the φ_j are the same for all mappings of the class. The interval on which the inequalities (2) hold is also independent of mappings, i.e. for the classes $\widetilde{BL}_k$ and BL_k (under some additional natural normality conditions) the inequalities are of equicontinuous character.

The inequalities (2) allow us to estimate for these classes the orders of uniform equicontinuity and openness in closed domains (with respect to the relative metrics) and the orders of uniform continuity and openness in the interior of domains (this time in terms of the spherical or Euclidean metric), and to estimate the distortion of the level sets of the distances r_A and r to the boundary of the domain and the distortion of the area of the boundary annuli, boundary arcs, etc. (see [10] and [11]). The classes $\widetilde{BL}_k$ and BL_k contain conformal, quasiconformal, and more general mappings; moreover, the results obtained turn out to be new even for conformal mappings.

The extension of the main results to the case of finitely connected domains is contained in Lukin's article [15], and the extension to the case of domains of arbitrary connectedness is contained in a paper of V. P. Luferenko and V. M. Miklyukov, which is being prepared for publication.*

An analogous investigation was conducted in the case where the compacta are bounded parts of the Lobachevsky plane, r and ρ are the hyperbolic metrics, and r_A and ρ_B are introduced on the basis of these metrics (Baklanov and Suvorov [16]). As a realization of the scheme of §1 we consider a general class of planar mappings such that the boundary correspondence has a more general character (not relative to the prime ends—see Baklanov [17]).

A general class of mappings transforming the entire finite plane into the same plane (a case excluded in other articles) has been considered by Ustinov in [18].

We have described the results for the cases of planar mappings of the classes BL_k and $\widetilde{BL}_k$, i.e. for the case $\alpha = 1$. For $\alpha > 1$ the classes $BL_k^\alpha(1) \subset BL_k$ and $BL_k^\alpha(\mu) \subset \widetilde{BL}_k$, $\mu = (1 + f_1^2 + f_2^2)^{-1}$, are smaller. As follows from Sobolev's imbedding theorems mappings of these classes of domains, with sufficiently smooth boundaries, can be extended by continuity to the boundary, and in closed (relative to the Euclidean metric) domains the upper estimate (2) holds with $\varphi_2(t) \equiv kt^\beta$, where $0 < \beta \leq 1$ and β depends on k. Consequently, if a mapping is a homeomorphism, then every prime end of the image domain always contains only one point. This means that two arbitrary simply connected domains cannot be mapped onto one another within this class. Other properties specific to these classes are not known.

For $0 < \alpha < 1$ the classes $BL_k^\alpha(1) \supset BL_k$ and $BL_k^\alpha(\mu) \supset \widetilde{BL}_k$ can essentially be extended. The a priori introduction of boundary elements in admissible domains is impossible, as shown by Ovchinnikov's example. Inequalities of type (2), equipotent relative to a class, are impossible; the equicontinuity of the class does not hold. Further properties of these classes have not been studied.

In the case $\alpha = 1$, Ovchinnikov has shown by example that the inequalities (2) with the φ_j indicated above are order-sharp. Obviously the inequalities are also sharp for the subclass of conformal mappings of simply connected domains with arbitrary boundaries, although we do not know of appropriate examples.

Let us note one more generalization of inequality (4) obtained by V. M. Miklyukov. Let $w = T(z)$ be a continuous mapping of a domain D having generalized partial derivatives of first order which are square summable over any compactum

---

**Editor's note.* This paper seems never to have appeared.

$F \subset D$. Assume that there exists a twice continuously differentiable nonnegative real function $\Phi(x)$ which is convex downwards (i.e. $\Phi''(x) > 0$) such that

$$\iint\limits_D \Phi\left[\sqrt{\lambda(z,T)}\right] dx\,dy < \infty.$$

Then the inequality

$$\int_{\tau_1}^{\tau_2} \Phi\left[\frac{L[T(\Gamma_t)]}{2\pi t}\right] t\,dt \le \iint\limits_{D_{\tau_1\tau_2}} \Phi\left[\sqrt{\lambda(z,T)}\right] dx\,dy, \tag{6}$$

holds for the mappings of this class, where L is the usual length and the remaining notation is the same as above.

For $\Phi(x) \equiv x^2$ we obtain the class BL, and (6) transforms into (4).

If the integral $\int_0^a \Phi(1/r) r\,dr$ diverges, then there exist functions φ_1 and φ_2 satisfying the conditions of §1 such that the estimates (1) and (2) of the distortion of the relative distances under homeomorphisms of the indicated class hold with these functions, where φ_2 is now not necessarily expressed in terms of logarithms. With the help of these estimates the results established for the classes BL and BL_k with the help of inequality (2) can also be extended to this larger class of mappings.

We know that even for a Q-quasiconformal mapping of a disc onto a disc the extension to the boundary need not be absolutely continuous on the boundary circle. Therefore it is interesting to find for these and for more general mappings sets of prime ends or some other such boundary formations of a domain which could be transformed under a mapping of the domain, e.g. onto a disc, into a set of Lebesgue measure zero on the circle. Two results for mappings $T \in BL$ with $T^{-1} \in BL$ (or for mappings of class $\widetilde{BL}$) containing sufficient conditions have been obtained by Suvorov [20], starting from inequality (4), and by Kufarev [21], with the help of inequality (3). However, this question cannot be considered as settled.

Most of the results that have been obtained reveal, in our opinion, basic metric properties of mappings. For example, the properties of equicontinuity and openness in the interior of a domain give (in combination with a general metric-topological theorem of [10]) sufficient conditions for the existence of a connection between the uniform convergence of a sequence of mappings in the interior of the domain and the convergence to the kernel of the sequence of the image domains [10], a connection completely analogous to the one which exists in the case of conformal mappings (a theorem of Carathéodorory). One more example: the properties of uniform equicontinuity and openness of mappings in closed domains ensure the possibility of solving the problem of boundary correspondence under mappings of domains with variable boundaries. Namely, we are guaranteed the applicability of the theory of prime ends of a sequence of domains converging to a nonsingular kernel, first developed for conformal mappings [19], to mappings of class $\widetilde{BL}_k$ if $T^{-1} \in \widetilde{BL}_k$ in the correspondence domains [10].

We give here the main theorem on boundary correspondence under mappings of domains with variable boundaries. As a preliminary let us recall how the basic notion of the corresponding theory, the notion of prime end of a sequence of domains [10], is introduced.

Let a sequence $B(B_0) = \{B_n\}$ of simply connected plane domains B_n converge relative to the origin to a nonsingular kernel B_0 (in the sense of Carathéodory) whose boundary contains more than one point.

The sequence $\{\beta_n\} = \beta\{\beta_0\}$ of Jordan cross cuts lies over the cross cut β_0 of the kernel B_0 if $\lim \beta_n = \beta_0$ (topological limit) and the points $P' \in B_0$ and $P'' \in B_0$ separated by the cross cut β_0 in B_0 are also separated by β_n in B_n for $n > n(P', P'')$. Such a sequence determines (to within a finite number of terms) the sequences $b(b'_0) = \{b'_n\}$ and $b(b''_0) = \{b''_n\}$ of connected components of $B_n \setminus \beta_n$ containing P' and P'' respectively. These sequences are called the "elements" lying over the connected components b'_0 and b''_0 respectively of the set $B_0 \setminus \beta_0$. A sequence $\{b(b_0^k)\}$ of elements lying respectively over the members of the chain (in the sense of Carathéodory's theory of prime ends) of subdomains $b_0^1 \supset b_0^2 \supset \cdots$ of the kernel B_0 is called a *chain of elements.* A chain $\{b(b_0^k)\}$ occurs in the element $a(a_0)$ if for some k_1 and n_1 we have $b_n^{k_1} \subset a_n$ for all $n \geq n_1$, and it occurs in the chain $\{a(a_0^k)\}$ if it occurs in every $a(a_0^k)$.

Two chains occurring in one another are considered as equivalent, and an equivalence class of chains is called an *end* of the sequence of domains $B(B_0)$. The notion of divisibility of ends and the notion of a prime end of $B(B_0)$ are now introduced according to Carathéodory. A one-to-one correspondence between the prime ends of B_0 and the prime ends of $B(B_0)$ can be established in a natural way so that the prime ends of $B(B_0)$ can be denoted as $e_b(e_{b_0})$, where e_{b_0} is the corresponding prime end of the kernel B_0.

We shall call the pair consisting of an interior point $e \in B_n$ (or a prime end e of the domain B_n) and the number n an *associated point* $(e, n) \equiv e_n$ of the domain B_n; $(e, n) = (e', n')$ if and only if $e = e'$ and $n = n'$. A sequence of associated points is said to be *regular* if it does not contain more than one prime end (or one interior point) of each of the domains of the sequence $B(B_0)$.

A regular sequence $\{e_{n_i}\}$ converges to a prime end $e_b(e_{b_0})$ if e_{n_i} occurs in every element of the chain $\{b(b_0^k)\}$ that determines $e_b(e_{b_0})$, i.e. if starting from a certain $i(k)$ every e_{n_i} occurs in b_n^k, where $b_{n_i}^k$ is one of the domains forming $b(b_0^k)$.

It is proved that from every regular sequence we can choose a subsequence converging to a prime end of $B(B_0)$ or to a point of the kernel B_0.

The principal result for the case of mappings of class $\widetilde{BL}_k$ is formulated as follows.

We consider a sequence $A(A_0) = \{A_n\}$ $(B(B_0) = \{B_n\})$ $(n = 1, 2, \ldots)$ of simply connected domains of the z-plane (of the w-plane), not containing the point at infinity in their interior and converging to a nonsingular kernel $A_0 \ni O$ (to $B_0 \ni O$). It is assumed that we can find a number $\gamma_0 > 0$ such that the spherical diameters in the metric ($\frac{1}{2}$) of the boundaries of all the domains are not less than γ_0.

Let

$$w = T_n(z), \quad T_n(O) = O \qquad (n = 1, 2, \ldots),$$

be a sequence of homeomorphisms of the domains, closed by prime ends, of the sequence $A(A_0)$ onto the domains of the sequence $B(B_0)$, $\tilde{B}_n = T_n(\tilde{A}_n)$, where $T_n \in \widetilde{BL}_k$ in A_n, $T_n^{-1} \in \widetilde{BL}_k$ in B_n, and k does not depend on n.

It is also assumed that the sequence $\{T_n\}$ (the sequence $\{T_n^{-1}\}$) converges uniformly in the interior of the domain A_0 (of B_0) to the homeomorphism T_0 (T_0^{-1}) of A_0 onto B_0 (of B_0 onto A_0).

Then there exists a one-to-one correspondence between the prime ends of the sequence $A(A_0)$ and the prime ends of the sequence $B(B_0)$ which is established by the limit mapping T_0, $\tilde{B}_0 = T_0(\tilde{A}_0)$, and which has the following properties.

If infinite sequences $\{e^z_{n_i}\}$ ($e^z_{n_i}$ is an interior point or a prime end of the domain A_{n_i}) and $\{e^w_{n_i}\}$ ($e^w_{n_i}$ is an interior point or a prime end of B_{n_i}) are connected by the equalities

$$T_{n_i}(e^z_{n_i}) = e^w_{n_i},$$

then from the convergence of $\{e^z_{n_i}\}$ to a prime end $e_a(e_{a_0})$ of the sequence $A(A_0)$ follows the convergence of $\{e^w_{n_i}\}$ to the prime end $e_b(e_{b_0})$ of $B(B_0)$, where $e_{b_0} = T_0(e_{a_0})$, and conversely, from the convergence of $\{e^w_{n_i}\}$ to $e_b(e_{b_0})$ follows the convergence of $\{e^z_{n_i}\}$ to $e_a(e_{a_0})$, where $e_{a_0} = T_0^{-1}(e_{b_0})$.

Let us observe that the topological content of the theory of prime ends of a sequence $B(B_0)$ reduces to the introduction of a boundary in the set B of all the associated points of the sequence $B(B_0)$ (which can be turned into a topological space) by two equivalent methods—with the help of certain families of open sets of associated points (this ensures geometrical visualization and ease of application), or with the help of certain families of sequences of associated points (which allows us to use the theory in studying the properties of sequences of mappings). The points of the kernel B_0 and the prime ends of the sequence $B(B_0)$ are the boundary elements of the topological space B; their totality is the boundary of B "in the restricted sense" of Myshkis [4], [5], and on the whole the construction is completely analogous to Carathéodory's construction (the theory of prime ends). The above cited theorem establishes a homeomorphism between the completed spaces $\tilde{A}$ and $\tilde{B}$ under the mappings $\{T_n\}$. In [10] numerous applications of this theorem to convergence problems for sequences of schlicht mappings are given. Moreover, in [10] and other articles a purely topological theory of prime ends of sequences of domains is systematically developed; this theory turns out to be quite interesting.

Here is one more very recent result of Suvorov [35], establishing, by means of the above-mentioned main theorem, the equipotent stability of conformal mappings of closed domains in the class $\widetilde{BL}_k$.

We consider a family $\{T\}$ of planar homeomorphisms of simply connected closed (by Carathéodory's prime ends) domains $\{\tilde{D}_T\}$ onto closed domains $\{\tilde{\Delta}_T\}$

such that $T(\tilde{D}_T) = \tilde{\Delta}_T$, where $T(O) = O$ and $T, T^{-1} \in \widetilde{BL}_k$. For every $T \equiv U + iV \in \{T\}$, let

$$\iint\limits_{D_T} \frac{(U_z \mp V_y)^2 + (U_y \pm V_x)^2}{(1+|T|^2)^2}\, dx\, dy \leq \varepsilon,$$

where the upper or lower signs are chosen according as T preserves or changes the orientation. The families $\{D_T\}$ and $\{\Delta_T\}$ have nonsingular kernels relative to the point O, and the spherical diameters of the boundaries of all the domains are bounded below by the same positive number.

In every domain D_T we choose a prime end e^{D_T} and consider a family $\{\Phi_T\}$ of conformal mappings of the first or second kind oriented in the same way as the corresponding T and such that $\Phi_T(\tilde{D}_T) = \Delta_T$, $\Phi_T(0) = 0$, and $\Phi_T(e^{D_T}) = T(e^{D_T})$. Then there exist functions $\lambda_j(\varepsilon)$, $\lambda_j \to 0$ as $\varepsilon \to 0$ $(j = 1, 2)$, such that for every $T \in \{T\}$ we have

$$\rho\left[T(e), \Phi_T(e); \tilde{\Delta}_T; \tfrac{1}{2}\right] \leq \lambda_1(\varepsilon), \qquad \rho\left[T^{-1}(e), \Phi_{T^{-1}}(e); \tilde{D}_T; \tfrac{1}{2}\right] \leq \lambda_2(\varepsilon),$$

where the two inequalities hold in the closed domains $\tilde{D}_T$ and $\tilde{\Delta}_T$ respectively.

The method of investigation in the planar case, and also the problems solved here, serve as a model for the search for analogues in the case of spatial mappings. This method is also suitable for the study of nonschlicht planar mappings, e.g., monotonic or interior mappings. The possibilities available here are far from exhausted.

§5. The case $n \geq 3$

The basic inequality of the method is inequality (5). Initially the case $n = p = 3$ was considered by Ovchinnikov and Suvorov [13]. The results they obtained, with new additions, were also carried over to the case $n > 3$ by Ovchinnikov [14]. Let us briefly list these results in the generality which Ovchinnikov gave them.

An upper bound was obtained for the distortion of the Euclidean metric in the interior of arbitrary domains for monotone mappings of class $BL^{n/2}$. For the class $BL_k^{n/2}$ this bound has an equipotent character; hence the order of equicontinuity of this class of mappings is determined in the interior of the domain. Problems on the compactness of the class are considered; in particular, its normality is established. It was proved that there do not exist nonconstant montone mappings in the class $BL^{n/2}$ which are defined on all of E^n (an analogue of Liouville's theorem) and that isolated singularities of homeomorphisms of this class are always of the removable type so that, for example, an n-dimensional ball with deleted center cannot be homeomorphically mapped onto a domain of E^n whose boundary consists of two components which are different from singletons. The order of growth of the mappings is indicated for the case of mappings of a sphere.

In the course of realizing the construction of §1 the following is established. Let D be a domain of E^n containing the origin O. By the *relative* (in $D \setminus O$)

distance between the points $x', x'' \in D \setminus O$ we mean the number

$$\rho(x', x''; D \setminus O) = \inf d(K), \tag{7}$$

where d is the Euclidean diameter of the set, the infimum being taken over all possible relative continua $K \subset D \setminus O$ satisfying the conditions a) $\overline{K} \not\ni O$ and b) $K \ni (x' \cup x'')$; or, if $K \not\ni (x' \cup x'')$, then $\overline{K} \cap \partial D$ is not empty and O belongs to that component of $D \setminus K$ which does not contain x' and x''.

The relative metric (7) turns out to be a metric. The completion of the metric space $D \setminus O$ relative to the metric (7) adjoins the point O and the "boundary elements" which are equivalence classes of Cauchy sequences in $D \setminus O$ not converging in $D \setminus O$. Let $\widetilde{\partial D}$ be the set of these boundary elements; in the usual manner, $\tilde{D} = D \cup \widetilde{\partial D}$ can be turned into a complete (but not necessarily compact) metric space with the metric $\rho(e_1, e_2; \tilde{D})$, which in the case of a planar simply connected bounded domain numerically coincides with the relative metric of §4.

The following theorem has been proved: If $y = T(x)$ is a homeomorphism of class $BL^{n/2}$ of the ball D: $|x| < 1$ onto a domain Δ such that $T(O) = O$, then T can be extended to a mapping continuous on $\overline{D}$ such that $T(\overline{D}) = \tilde{\Delta}$ and to each point $x \in \overline{D} \setminus D$ there corresponds, under the mapping T, a unique boundary element of $\tilde{\Delta}$, where, if x' and x'' are any points of $\overline{D}$ such that $0 < |x' - x''| < \frac{1}{2}$, then

$$\rho[T(x'), T(x''); \tilde{\Delta}] < \varphi_2(|x' - x''|),$$

where

$$\varphi_2(\alpha) = \left[4M_n \int_D \lambda^{n/2}(x, T)\, dx\right]^{1/n} \left(\ln \frac{1}{2\alpha}\right)^{-1/2}. \tag{8}$$

Consequently, for mappings of class $BL_k^{n/2}$ the inequality has an equipotent character. Hence a necessary condition for the possibility of mapping a ball onto a domain with the help of a homeomorphism of class $BL^{n/2}$ is the compactness of the completion of the domain with respect to the relative metric. This simple observation makes it easy for us to construct examples of domains homeomorphic to a ball, onto which the ball cannot be mapped by a homeomorphism of class $BL^{n/2}$ (hence neither by a Q-quasiconformal mapping).

A consequence of (8) is also a two-sided estimate (of the type of (1) and (2)) of the distortion of the Euclidean metric under mappings of two balls, one onto the other, such that the homeomorphism T, $T^{-1} \in BL^{n/2}$, of a ball onto a ball can be extended to a homeomorphism of closed balls.

In [13] two examples (due to I. S. Ovchinnikov and V. M. Miklyukov) are given. The first example shows the impossibility of a nontrivial lower estimate of the distortion of the relative metric under the homeomorphisms of class $BL^{n/2}$ of closed Jordan domains, and the second example shows the impossibility of an upper estimate in the same general situation. Consequently, the above-mentioned result about the homeomorphism of balls cannot be extended even to homeomorphisms of Jordan domains. These examples also show that such homeomorphisms of class $BL^{n/2}$ do not realize a boundary correspondence relative to points. This fact brings out the essential difference between the spatial case and the planar case, and also that between the cases of spatial mappings of class $BL^{n/2}$ and spatial Q-quasiconformal mappings. Moreover, Miklyukov has constructed an example of two homeomorphisms of class $BL^{3/2}$ of a ball onto a single Jordan domain Δ such that the complete preimage of a point $y_0 \in \partial\Delta$ is in one case a single point, and in the other case a whole continuum on the boundary of the ball. This example is essential, for it shows that one cannot introduce boundary elements a priori in admissible domains; it also brings out the fact that the results in [13] relating to the elucidation of the general character of the boundary correspondence under homeomorphisms of class $BL^{3/2}$ are definitive in a certain sense.

The same examples show that the class $BL^{n/2}$ is larger than the class of mappings considered by V. A. Zorich (a student of B. V. Shabat), who completely solved the problem of boundary correspondence for spatial Q-quasiconformal mappings of a ball onto a domain (see [24]–[26]).

As in the planar case, we can introduce prime ends in domains of E^n in analogy with Carathéodory's theory in the planar case. However, it is immediately clear that the completion of a domain by these boundary elements is in general not compact, and this is why Carathéodory's scheme was rejected in the spatial case (Kaufman [27]). Ovchinnikov in [23] and [28] nevertheless carried out this construction and showed that for a certain class of domains (which contains, for example, all simply connected bounded domains that are compact with respect to the relative metric) this completion turns out to be compact, and that for this class of domains the prime ends can be identified with the boundary elements with respect to the relative metric. In contrast to the planar case it turns out that in domains which are homeomorphic to a ball and compact with respect to the relative metric there may exist prime ends of a "fifth type" (Carathéodory has four types of prime ends in the planar case) which have a disconnected closed set of principal points. However, a domain having such prime ends cannot be the image of a ball under a homeomorphism of class $BL^{n/2}$, since Ovchinnikov [29] has shown that under such homeomorphisms the limit set relative to any nontangential path in the ball approaching a boundary point of A coincides with the set of principal points of the prime end, $e = T(A)$ (an analogue of Lindelöf's theorem).

The class of domains having prime ends of the fifth type is rather large. For example, if we consider an arbitrary closed disconnected set F on a face H of a cube, then we can construct a domain which is homeomorphic to a ball and compact with respect to the relative metric such that it contains a prime end whose body is the set H and the set of principal points of this prime end coincides with the set F.

Every homeomorphism of class $BL^{n/2}$ of a ball onto a domain can be extended to a continuous mapping of the closed ball such that to every boundary point of the ball there corresponds a unique prime end of the domain. If we assume in addition that the inverse mapping belongs to the class $BL^{n/2}$, then for $n = 2$ the boundary correspondence under the mapping would be one-to-one relative to the prime ends, but for $n \geq 3$, as is shown by examples, the boundary correspondence may now not be one-to-one. If we consider the smaller class of quasiconformal mappings of a ball, then the boundary correspondence, as in the planar case, would be a one-to-one correspondence relative to the prime ends (see [24] and [26]).

Other criteria have also been obtained for domains which are compact with respect to the relative metric, but are not images of a ball under homeomorphisms of class $BL^{n/2}$. For example, every domain having a prime end dividing the boundary of the domain and not having accessible points is such a domain (Zhukov and Miklyukov [30]). It has also been observed that if a prime end divides the boundary of a domain into two components, then the possibility or impossibility of a homeomorphism T of a ball onto such a domain, where $T, T^{-1} \in BL^{3/2}$, depends on the order of contact of these components (Lapko and Ovchinnikov [31]).

For monotone mappings of class $BL^{3/2}$ with continuous derivatives of first order, Miklyukov in [32] has proved that if a mapping T of a ball has a limit value at a boundary point along a certain path, then it has this value as the angle limit value, and that T has almost everywhere on the boundary of the ball finite angle limit values. In [30] the second result has been strengthened: An analogous conclusion has been established for monotone mappings of class $BL^\alpha(1)$ $(\alpha > 1)$ of a three-dimensional ball in E^3.

We have already mentioned above that for homeomorphisms of class $BL^{n/2}$ the construction of §1 cannot be realized if it aims at obtaining the two-sided estimate (2) so that the question of the boundary correspondence can be meaningfully considered, since the latter question cannot, in general, be solved along the lines of §1. Many results of a metric character (e.g., the analogue of Carathéodory's theorem; see §1) can be obtained only in the presence of two-sided estimates. However there remains a certain arbitrariness in the choice of the relative metric. Ovchinnikov has taken advantage of this fact to define a new relative metric (see [33] and [34]), the completion relative to which turns an arbitrary domain into a compactum. Let us give his definition:

Let $x', x'' \in D$ and $D \ni O$. We put

$$\sigma(x', x''; D) = \min\left[\rho(x', x''), \inf \sup_{x \in F} \rho(x, \rho D)\right],$$

where ρ is the Euclidean metric, the infimum being taken over all sets F which are closed relative to D, separate the points x' and x'' from O in D, and are such that, if $\overline{F} \cap \partial D$ is empty, then F also separates $x' \cup x''$ from ∂D in $\overline{D}$ (∂D is assumed to be nonempty). σ is a metric. Let $\tilde{D}$ be the completion of D relative to σ. Then $\tilde{D} \setminus D = \widetilde{\partial D}$ turns out to be a single point in the metric σ, and this "point" can be identified with the usual boundary ∂D. However σ satisfies all the requirements of §1, and in contrast to the above-mentioned metric $\rho(x', x'', \tilde{D})$ it is possible to obtain estimates of the type (2) for the new metric with

$$\varphi_2(\alpha) = \left[4M_n \int_D \lambda^{n/2}(x,T)\,dx\right]^{1/n} \left(\ln\frac{1}{\alpha}\right)^{-1/2} \quad \text{and} \quad \varphi_1 = \varphi_2^{-1},$$

where the estimates are equipotent for mappings of class $BL_k^{n/2}$.

Of course, a conclusion about the boundary correspondence under mappings which follows from these estimates (more precisely, from (1)) is quite trivial: The ordinary boundary transforms into the ordinary boundary, but since estimate (2) turns out to be true up to the boundary, we can estimate the rate of approach of a point-image to the boundary of the domain as depending on the rate of approach of the point-preimage to the boundary of the domain. This allows us to establish a number of substantial metric properties of the classes $BL^{n/2}$ and $BL_k^{n/2}$ following the methods of the planar case. We obtain two-sided estimates of the distortion of level sets with respect to the relative and the Euclidean metrics (equipotent in the class $BL_k^{n/2}$), and establish a two-sided estimate of the distortion of the Euclidean metric in the interior of a domain under mappings of arbitrary domains, whose consequences are the properties of equicontinuity and equiopenness (relative to the origin when $T(O) = O$) in the sense of the definitions in [10], an analogue of a theorem of Carathéodory on mappings of domains with variable boundaries.

§6. Concluding remarks

The articles described by us can be divided into three groups, depending on the three possible topological situations: the ambient compacta, the domains under consideration and the relative metrics are such that: 1) the completion $\tilde{A}$ is a compactum and $\widetilde{\partial A} = \tilde{A} \setminus A$ does not reduce to a single point; or 2) $\tilde{A}$ is a compactum, and $\widetilde{\partial A}$ is a singleton; or 3) $\tilde{A}$ is in general not compact, but $\widetilde{\partial A}$ contains more than one point. It should be observed that the use of a situation is determined by the essence of the matter, so that these situations are not mutually interchangeable if we remain at the standpoint of a priori introduction of the boundary elements in the domains in a unique manner for all mappings of the selected classes.

In Kaufmann's paper [27], boundary elements, i.e. prime ends, are introduced in spatial domains, and the completion of a domain by such boundary elements is compact. Unfortunately, however, till now we have not found a class of mappings for which the boundary correspondence can be realized relative to Kaufmann's prime ends. Therefore the results of §5 remain the most general.

References

1. C. Carathéodory, *Über die Begrenzung einfach zusammenhängender Gebiete*, Math. Ann. **73** (1913), 323–370.

2. S. Mazurkiewicz, *Über die Definition der Primenden*, Fund. Math. **26** (1936), 272–279.

3. M. Lavrentieff [Lavrent′ev], *Sur une classe de représentations continues*, Mat. Sb. **42** (1935), 407–423.

4. A. D. Myshkis, *On the concept of boundary*, Mat. Sb. **25(67)** (1949), 387–414; English transl. in Amer. Math. Soc. Transl. (1) **8** (1962).

5. ——, *On the equivalence of certain methods of defining a boundary*, Mat. Sb. **26(68)** (1950), 228–236; English transl. in Amer. Math. Soc. Transl. (1) **8** (1962).

6. M. V. Baklanov and G. D. Suvorov, *On the closure of certain classes of mappings with respect to uniform convergence*, Trudy Tomsk. Gos. Univ. Ser. Mekh.-Mat. **182** (1965), vyp. 3, 3–14. (Russian)

7. V. K. Ionin and G. D. Suvorov, *On the components of the level sets of the function: distance to a plane continuum*, Dokl. Akad. Nauk SSSR **129** (1959), 496–498. (Russian)

8. B. P. Kufarev and N. G. Nikulina, *Lebesgue measure of subsets of Euclidean space as the leading variation of the function defined by the distance to a closed set*, Dokl. Akad. Nauk SSSR **160** (1965), 1004–1006; English transl. in Soviet Math. Dokl. **6** (1965).

9. Jacqueline Lelong-Ferrand, *Représentation conforme et transformations à intégrale de Dirichlet bornée*, Gauther-Villars, Paris, 1955.

10. G. D. Suvorov, *Families of plane topological mappings*, Izdat. Sibirsk. Otdel. Akad. Nauk SSSR, Novosibirsk, 1965. (Russian)

11. ——, *Metric properties of plane univalent mappings of closed domains*, Trudy Tomsk. Gos. Univ. Ser. Mekh.-Mat. **182** (1965), vyp. 3, 46–58. (Russian)

12. B. P. Kufarev, *Relations of the type of the "length and area principle"*, Dokl. Akad. Nauk SSSR **170**(1966), 268–270; English transl. in Soviet Math. Dokl. **7** (1966).

13. I. S. Ovchinnikov and G. D. Suvorov, *Transformations of the Dirichlet integral and space mappings*, Sibirsk. Mat. Zh. **6** (1965), 1292–1314. (Russian)

14. I. S. Ovchinnikov, *A "length and area principle" type inequality for n-dimensional space*, Trudy Tomsk. Gos. Univ. Ser. Mekh.-Mat. **189** (1966), 86–95. (Russian)

15. V. N. Lukin, *Metric properties of a certain class of mappings of finitely connected plane domains*, Trudy Tomsk. Gos. Univ. Ser. Mekh.-Mat. **189** (1966), 66–79. (Russian)

16. M. V. Baklanov and G. D. Suvorov, *The distortion of relative distances in closed domains under topological mappings of class $BL(\Lambda)$ with respect to a conformally invariant metric*, Trudy Tomsk. Gos. Univ. Ser. Mekh.-Mat. **182** (1965), vyp. 3, 15–26. (Russian)

17. M. V. Baklanov, *Some classes of topological mappings of plane domains with the boundary correspondence not according to prime ends*, Trudy Tomsk. Gos. Univ. Ser. Mekh.-Mat. **189** (1966), 3–12. (Russian)

18. Yu. K. Ustinov, *On a certain class of plane mappings*, Trudy Tomsk. Gos. Univ. Ser. Mekh.-Mat. **189** (1966), 104–110. (Russian)

19. G. D. Suvorov, *On the prime ends of a sequence of plane domains converging to a nucleus*, Mat. Sb. **33(75)** (1953), 73–100; English transl. in Amer. Math. Soc. Transl. (2) **1** (1955).

20. ——, *Univalent mappings of plane domains and sets of prime ends of generalized measure zero*, Dokl. Akad. Nauk SSSR **152** (1963), 296–298; English transl. in Soviet Math. Dokl. **4** (1963).

21. B. P. Kufarev, *Zero sets and homeomorphisms with finite Dirichlet integral*, Dokl. Akad. Nauk SSSR **173** (1967), 1257–1259; English transl. in Soviet Math. Dokl. **8** (1967).

22. ——, *On the theory of plane mappings of certain classes*, Candidate's Dissertation, Tomsk State Univ., Tomsk, 1965. (Russian)

23. I. S. Ovchinnikov, *Some metric properties of space mappings*, Candidate's Dissertation, Tomsk State Univ., Tomsk, 1966. (Russian)

24. V. A. Zorich, *On a boundary correspondence for Q-quasiconformal mappings of a sphere*, Dokl. Akad. Nauk SSSR **145** (1962), 31–34; English transl. in Soviet Math. Dokl. **3** (1962).

25. ——, *Boundary properties of a class of mappings in a space*, Dokl. Akad. Nauk SSSR **153** (1963), 23–26; English transl. in Soviet Math. Dokl. **4** (1963).

26. ——, *Determination of boundary elements by means of sections*, Dokl. Akad. Nauk SSSR **164** (1965), 736–739; English transl. in Soviet Math. Dokl. **6** (1965).

27. B. Kaufmann, *Über die Berandung ebener und räumlicher Gebiete*, Math. Ann. **103** (1930), 70–144.

28. I. S. Ovchinnikov, *Prime ends of a certain class of space regions*, Trudy Tomsk. Gos. Univ. Ser. Mekh.-Mat. **189** (1966), 96–103.(Russian)

29. ——, *On the nonexistence of homeomorphic mappings from a certain class taking a ball onto a domain having prime ends with a disconnected set of principal points*, Trudy Tomsk. Gos. Univ. Ser. Mekh.-Mat. (to appear). (Russian)

30. V. A. Zhukov and V. M. Miklyukov, *Angular boundary values of space mappings*, Trudy Tomsk. Gos. Univ. Ser. Mekh.-Mat. **200** (1968), 88–95. (Russian)

31. A. M. Lapko and I. S. Ovchinnikov, *Nonexistence of a certain class of mappings of a sphere onto a domain*, Trudy Tomsk. Gos. Univ. Ser. Mekh.-Mat. **200** (1968), 165–172. (Russian)

32. V. M. Miklyukov, *Boundary properties of a class of mappings in space*, Trudy Tomsk. Gos. Univ. Ser. Mekh.-Mat. **189** (1966), 80–85. (Russian)

33. I. S. Ovchinnikov, *Metric properties of mappings in the class* $BL^{3/2}$, Dokl. Akad. Nauk SSSR **161** (1965), 526–529; English transl. in Soviet Math. Dokl. **6** (1965).

34. ——, *Metric properties of mappings in the class* $BL^{3/2}$, Trudy Tomsk. Gos. Univ. Ser. Mekh.-Mat. **182** (1965), vyp. 3, 32–45. (Russian)

35. G. D. Suvorov, *Uniform stability of conformal mappings of closed domains*, Ukrain. Mat. Zh. **20** (1968), 78–84; English transl. in Ukrainian Math. J. **20** (1968).

36.* V. P. Luferenko and G. D. Suvorov, *Families of homeomorphisms uniformly equicontinuous in relative metrics*, Dokl. Akad. Nauk SSSR **192** (1970), 30–33; English transl. in Soviet Math. Dokl. **11** (1970), No. 3.

Translated by Y. K. CHOUDHARY

*Added in translation.

Amer. Math. Soc. Transl.
(2) Vol. **134**, 1987

On Noetherian Spaces

V. I. MALYKHIN

I. Recall (see [1]) that a family of sets is said to be (weakly) Noetherian if each nonempty element of it is contained in only finitely (countably) many elements of the family. A base or π-base for a topological space is said to be (weakly) Noetherian if it is a (weakly) Noetherian family. A space with such a base is named accordingly [1].

Spaces having Noetherian and weakly Noetherian bases and π-bases were investigated in [2]–[4], and several interesting assertions about such spaces were proved. However, the answers to the following questions remained unclear.

1. Do there exist (weakly) Noetherian compact Hausdorff spaces that do not satisfy the Suslin condition?

2. Do there exist weakly Noetherian compact Hausdorff spaces that are not Noetherian?

3. Can an extremally disconnected compact Hausdorff space be Noetherian?

4. Does a compact Hausdorff space with the Suslin condition or even a countable space necessarily have a weakly Noetherian π-base?

Below we give positive answers to questions 1–3 (under the assumption of the continuum hypothesis CH for the second question) and a negative answer to the fourth question under the assumption of Martin's axiom MA (see [5] and [6] for the basic facts about Martin's axiom).

II. LEMMA 1. *Suppose that the space X has a net of cardinality at most $w(X)$ composed of compact subsets, and that $\chi(x, X) = w(X)$ for each point x of the space. Then any base for the space decomposes into a disjoint family of cardinality $w(X)$ whose members are bases.*

1980 *Mathematics Subject Classification* (1985 *Revision*). Primary 54A25; Secondary 03E50, 04A30.

Translation of Seminar on General Topology (P. S. Aleksandrov, editor), Moskov. Gos. Univ., Moscow, 1981, pp. 51–59; MR **83f:** 54023..

PROOF. Let $\mathcal{B}$ be some base for X, and let $\mathcal{S}$ be some net of cardinality at most $m = w(X)$ composed of compact subsets. For each $S \in \mathcal{S}$ fix a family $\mathcal{V}(S)$ of cardinality at most m composed of neighborhoods of it and forming a base for it (this can be done because S is compact and $w(X) = m$). Let $\mathcal{P}$ be the set of all possible pairs $\langle S, V\rangle$ where $V \in \mathcal{V}(S)$. For each $\gamma \in m$ let $\mathcal{P}_\gamma$ be a copy of $\mathcal{P}$; then denote by $\mathcal{W}$ the union $\bigcup\{\mathcal{P}_\gamma : \gamma \in m\}$ and suppose that some well-ordering is given on $\mathcal{W}$: $\mathcal{W} = \{W_\alpha : \alpha \in m\}$.

We construct a one-to-one mapping φ of $\mathcal{W}$ into the family of finite subsets of the base $\mathcal{B}$ (i.e., $\varphi(W)$ is a finite family of elements of $\mathcal{B}$ for each $W \in \mathcal{W}$), such that if $W' \neq W''$ and $W', W'' \in \mathcal{W}$, then $\varphi(W') \cap \varphi(W'') = \varnothing$ (it is assumed that the base $\mathcal{B}$ does not contain identical elements!) and, moreover, $S \subset \bigcup \varphi(\langle S, V\rangle) \subset V$ for every pair $\langle S, V\rangle$ in $\mathcal{W}$.

Assume that for some $\alpha \in m$ and every $\beta \in \alpha$ the mapping φ is already defined for W_β with the necessary conditions satisfied; in particular, assume that $|\varphi(W)_\beta)| < \aleph_0$. Consider the pair $W_\alpha = \langle S, V\rangle$. Since

$$|\bigcup\{\varphi(W_\beta : \beta \in \alpha\}| < m,$$

for each point $s \in S$ there is an element $B(s) \in \mathcal{B}$ such that $s \in B(s)$, $B(s) \subset V$, and $B(s) \notin \bigcup\{\varphi(W_\beta) : \beta \in \alpha\}$ (after all, $\chi(s, X) = m$). From the covering $\{B(s) : s \in S\}$ of S we select a finite subcovering $\{B_1, \dots, B_n\}$ and let $\varphi(W_\alpha) = \{B_1, \dots, B_n\}$. All the induction hypotheses are obviously satisfied, so it is possible to continue the process and construct a mapping φ with the required properties.

We now prove that for each $\gamma \in m$ the family $\bigcup \varphi(\mathcal{P}_\gamma) = \bigcup\{\varphi(W) : W \in \mathcal{P}_\gamma\}$ is a base for X. If Q is an open set, then Q is the union of some family Δ of sets in the net $\mathcal{S}$. For each $S \in \Delta$ there exists a neighborhood $V_0(S)$ contained in Q and such that $\langle S, V_0(S)\rangle \in \mathcal{P}_\gamma$. It is easy to see that $\bigcup\bigcup\{\varphi\langle S, V_0(S)\rangle : S \in \Delta, \langle S, V_0(S)\rangle \in \mathcal{P}_\gamma\} = Q$; consequently, any open set in the space is a union of elements in the family $\bigcup \varphi(P_\gamma)$, i.e., $\bigcup \varphi(P_\gamma)$ is a base for the space.

Recall that the index of compactness $\mathrm{ic}(X)$ of the space X is the minimum of the cardinal numbers m such that any open covering of the space contains a subcovering of cardinality at most m.

LEMMA 1′. *Suppose that X has a base $\mathcal{B}$ such that $\chi(x, X) = |\mathcal{B}|$ for every point of the space, and also $\sup\{\mathrm{ic}(\overline{B}) : B \in \overline{\mathcal{B}}\} < |\mathcal{B}|$. Then for any base $\mathcal{N}$ for the space there exists a disjoint family $(\mathcal{B}, \mathcal{N})$ of cardinality $|\mathcal{B}|$ whose members are bases for the space and a one-to-one mapping θ from $\bigcup(\mathcal{B}, \mathcal{N})$ to $\mathcal{N}$ such that $A = \theta(A) \cap B$, where $B \in \mathcal{B}$, for every element A in $\bigcup(\mathcal{B}, \mathcal{N})$ (i.e., every element in $\bigcup(\mathcal{B}, \mathcal{N})$ is the intersection of some element of $\mathcal{N}$ and some element of $\mathcal{B}$, and different elements of $\bigcup(\mathcal{B}, \mathcal{N})$ are obtained from different elements of $\mathcal{N}$).*

The proof is analogous to that of Lemma 1. We make only the following remark.

REMARK. If $\mathcal{N}$ is a Noetherian family, then so is $\bigcup(\mathcal{B}, \mathcal{N})$. Indeed, suppose that some element A in $\bigcup(\mathcal{B}, \mathcal{N})$ is contained in an infinite subfamily Δ of elements of $\bigcup(\mathcal{B}, \mathcal{N})$. Since A is open and is assumed to be nonempty, it contains some element N in $\mathcal{N}$; but then $N \subset \bigcap \theta(\Delta)$, and $\theta(\Delta)$ is an infinite subfamily in $\mathcal{N}$, because θ is one-to-one: we have contradicted the fact that the family $\mathcal{N}$ is Noetherian.

THEOREM 1. *Suppose that* $\chi(x, X) = w(X)$ *for every point of the Noetherian space* X*, and that at least one of the following conditions holds:*

1) *The space has a net of cardinality at most* $w(X)$ *composed of compact subsets.*

2) *The space has a base* $\mathcal{B}$ *such that* $\operatorname{Sup}\{\operatorname{ic}(\overline{B}) : B \in \mathcal{B}\} < w(X)$.

Then the product $X \times Y$ *is a Noetherian space if* $w(Y) \leq w(X)$.

PROOF. Let $\mathcal{K}$ be some base of cardinality at most $w(X)$ in Y. Under either of condition 1) or 2) the space X has a disjoint family $\{\mathcal{B}_K : K \in \mathcal{K}\}$ of bases which we index by the elements in $\mathcal{K}$. It is important that $\Delta = \bigcup\{\mathcal{B}_K : K \in \mathcal{K}\}$ is a Noetherian family (see Lemma 1 and the remark after Lemma 1′).

Let $\mathcal{R} = \{V \times K : V \in \mathcal{B}_K,\ K \in \mathcal{K}\}$. We prove that $\mathcal{R}$ is a base in the product $X \times Y$ and is a Noetherian family.

Clearly, $\mathcal{P} = \{W \times K : W \in \Lambda,\ K \in \mathcal{K}\}$ is a base in $X \times Y$. But for every $K \in \mathcal{K}$ the family $\mathcal{B}_K$ is a base for X; hence, for a given "rectangle" $W \times K$ there exists a subfamily $\Delta(W, K)$ of $\mathcal{B}_k$ such that $W = \bigcup \Delta(W, K)$; but then

$$W \times K = \bigcup\{V \times K : V \in \Delta(W, K) \subset \mathcal{B}_K\}.$$

Consequently, any element in $\mathcal{P}$ can be obtained by taking a union of elements in $\mathcal{R}$, and $\mathcal{R}$ is thereby a base for $X \times Y$.

If now $\mathcal{E}$ is some infinite subfamily of $\mathcal{R}$, then $\mathcal{E}_1 = \{V : V \times K \in \mathcal{E}\}$ is also an infinite subfamily of the Noetherian family Δ; therefore, $\operatorname{int} \cap \mathcal{E}_1 = \varnothing$, and this implies (since the topology on $X \times Y$ is the product topology) that also $\operatorname{int} \cap \mathcal{E} = \varnothing$. It has been proved that $\mathcal{R}$ is a Noetherian family.

COROLLARY 1. *For every space (compact Hausdorff space) there exists a Noetherian space (compact Hausdorff space) that maps openly and perfectly onto it.*

Indeed, for the space X the product $X \times I^{w(X)}$ will be such a space, because the Tychonoff cube I^m satisfies all the conditions of Theorem 1: $\chi(x, I^m) = m = \omega(I^m)$ for any point $x \in I^m$, and I^m is a Noetherian space, being a product of Noetherian spaces [3].

COROLLARY 2. *There exist Noetherian compact Hausdorff spaces that do not satisfy the Suslin condition.*

Let X be some Hausdorff space that does not satisfy the Suslin condition; then $X \times I^{w(X)}$ is a Noetherian compact Hausdorff space which, according to Theorem 1, does not satisfy the Suslin condition.

III. THEOREM 2. *If* $|A| \geq \operatorname{Sup}\{w(X_a)\colon a \in A\}$, *then the space* $\Pi\{X_a\colon a \in A\}$ *is Noetherian.*(1)

PROOF. Fix some base $\mathcal{B}$ of cardinality $|A|$ in $Z = \Pi\{X_a\colon a \in A\}$ with the product topology. It is not hard to construct by transfinite induction a one-to-one mapping $\varphi\colon \mathcal{B} \longrightarrow A$ such that $\varphi(B) > \max \operatorname{Supp} B$ for any element $B \in \mathcal{B}$ (here $\operatorname{Supp} B = \{a \in A\colon \pi_a(B) \neq X_a\}$). Further, in each X_a fix two proper open sets V_a' and V_a'' with union the whole of X_a, and let

$$r'(B) = B \cap \pi^{-1}_{\varphi(B)}(V'_{\varphi(B)}), \qquad r''(B) = B \cap \pi^{-1}_{\varphi(B)}(V''_{\varphi(B)}).$$

Clearly, $\varphi(B) \in \operatorname{Supp} r'(B) \cap \operatorname{Supp} r''(B)$. We set

$$\mathcal{B}_r = \{r'(B)\colon B \in \mathcal{B}\} \cup \{r''(B)\colon B \in \mathcal{B}\}.$$

This is a base for Z, because $B = r'(B) \cap r''(B)$ for each $B \in \mathcal{B}$. Let us prove that it is Noetherian. Suppose that Δ is an infinite subsystem in $\mathcal{B}_r$. Then either

$$\Delta' = \Delta \cap \{r'(B)\colon B \in \mathcal{B}\} \quad \text{or} \quad \Delta'' = \Delta \cap \{r''(B)\colon B \in \mathcal{B}\}$$

is infinite. Suppose that Δ' is infinite. We prove that $\operatorname{Int} \bigcap \Delta' = \varnothing$. Indeed, assume that W is a nonempty open set in Z and $W \subset \bigcap \Delta'$; but then

$$\operatorname{Supp} W \supseteq \bigcup \{\operatorname{Supp} r'(B) \in \Delta \in \Delta'\},$$

and the last set is infinite by construction, so $\operatorname{Supp} W$ is infinite, which contradicts the fact that Z has the product topology. The theorem is proved.

COROLLARY. *For any space* X *the power* X^n *is a Noetherian space if* $n \geq w(X)$.

REMARK. Suppose that in the product of the family A of spaces the family $\mathcal{W}$ of all regular open sets has cardinality $|A|$. Then it contains a Noetherian subfamily $\mathcal{W}^*$ such that each element of $\mathcal{W}$ is a union of two elements in $\mathcal{W}^*$. This follows obviously from the proof of Theorem 2.

THEOREM 3. *Suppose that in the product* Z *of a family* A *of Hausdorff spaces the family of all regular open sets has cardinality* $|A|$. *Then the space* $\overset{\circ}{Z}$ *of maximal systems of open sets with the finite intersection property is a Noetherian space in the natural topology (and is extremally disconnected).*

PROOF. If V is an open subset of Z, then let $\overset{\circ}{V} = \{\xi \in \overset{\circ}{Z}\colon V \in \xi\}$. It is known that the family $\{\overset{\circ}{V}\colon V \in \mathcal{W}\}$, where $\mathcal{W}$ is the family of all regular open sets in Z, forms a base in $\overset{\circ}{Z}$. It is easy to see that the family $\{\overset{\circ}{V}\colon V \in \mathcal{W}^*\}$ is also a base in $\overset{\circ}{Z}$. Here $\mathcal{W}^*$ is the subfamily of $\mathcal{W}$ mentioned in the above remark. It is also obvious that the family $\{\overset{\circ}{V}\colon V \in \mathcal{W}\}$ is Noetherian. This proves the theorem.

(1)Here and subsequently, the topology on the spaces under consideration is assumed to satisfy certain very weak conditions.

COROLLARY 1. *Suppose that Z is the product of a family A of separable regular spaces. Then $\overset{\circ}{Z}$ is a Noetherian space if $|A|^{\aleph_0} = |A|$, and is not weakly Noetherian if $|A|^{\aleph_0} \neq |A|$ and at least one of the following conditions holds:*

1) $|A| > \mathfrak{c}$.
2) *The π-weight of each factor is at most $|A|$.*

PROOF. If $|A|^{\aleph_0} = |A|$, then the conditions of Theorem 3 hold. Indeed, since it is separable and regular, there are at most $c = 2^{\aleph_0}$ regular open sets. But in the product Z each regular closed (and, hence, each regular open) set "is determined" by a countable face (see [7]); therefore, in all there are $|A|^{\aleph_0} \cdot \mathfrak{c} = |A|^{\aleph_0} = |A|$ canonical open subsets of Z. On the other hand, it is obvious that the π-weight of $\overset{\circ}{Z}$ is equal to $|A|$ if either condition 1) or 2) holds, while the weight of $\overset{\circ}{Z}$, an extremally disconnected compact Hausdorff space, is always an $\aleph_0$-admissible number (see, for example, [8]), i.e., $w(\overset{\circ}{Z})^{\aleph_0} = w(\overset{\circ}{Z})$. Consequently, $w(\overset{\circ}{Z}) > \pi w(\overset{\circ}{Z})$, from which it follows that Z is not a weakly Noetherian space, since in a weakly Noetherian space these topological invariants coincide [4].

COROLLARY 2. *The assertion that the absolute of the generalized Cantor discontinuum $D^{\aleph_1}$ is a Noetherian compact Hausdorff space is equivalent to CH.*

Indeed, CH is equivalent to $\aleph_1^{\aleph_0} = \aleph_1$.

IV. We now turn to weakly Noetherian spaces.

THEOREM 4. *If the weight of a Lindelöf space X is equal to $\aleph_1$ and $\pi\chi(x, X) > \aleph_0$ for every point $x \in X$, then X is a weakly Noetherian space.*

PROOF. Let the elements of some base $\mathcal{B}$ of cardinality $\aleph_1$ be numbered by countable ordinals: $\mathcal{B} = \{B_\alpha : \alpha \in \omega_1\}$. We shall define a mapping φ by assigning to each $B \in \mathcal{B}$ a countable collection of open sets with union B.

Suppose that φ has already been defined for every B_β with $\beta \in \alpha$, where α is some countable ordinal. Consider the set $\overline{B}_\alpha$. Since $\pi\chi(x, X) > \aleph_0$ for each point $x \in X$, the countable family $\Delta_\alpha = \bigcup\{\varphi(B_\beta) : \beta \in \alpha\}$ does not form a π-base at any point $b \in \overline{B}_\alpha$, and thus each $b \in \overline{B}_\alpha$ has a neighborhood $V(b)$ not containing any elements of Δ_a. From the covering $\{V(b) : b \in \overline{B}_\alpha\}$ of $\overline{B}_\alpha$ we now select a countable subcovering $\{V_n : n \in \omega_0\}$ and let $\varphi(B_\alpha) = \{V_n \cap B_\alpha : n \in \omega_0\}$. In this way it is not hard to define a mapping φ with the required properties.

It is perfectly clear that the family $\mathcal{E} = \bigcup\{\varphi(B_\alpha) : \alpha \in \omega_1\}$ is a base. Further, if $V \in \mathcal{E}$ and $V \in \varphi(B_\alpha)$ for some countable ordinal α, then V can be contained only in elements of the family $\bigcup\{\varphi(B_\beta) : \beta \leq \alpha\}$, and in only countably many of them. Consequently, the system $\mathcal{E}$ is weakly Noetherian. The theorem is proved.

COROLLARY [CH]. *$\beta N \backslash N$ is a weakly Noetherian compact Hausdorff space.*

LEMMA 2. *In any π-base for $\beta N \backslash N$ there is a strictly decreasing family of elements that is of type ω_1.*

PROOF. The elements of the π-base are assumed to be open and closed. A strictly decreasing family is itself not hard to construct by transfinite induction, using the widely known fact that every nonempty G_δ-set in $\beta N\backslash N$ has a nonempty interior (see, for example, [7]).

The next assertion is proved similarly.

LEMMA 2′ [MA]. *In any π-base for $\beta N\backslash N$ there is a strictly decreasing family of elements that is of the type of the first ordinal of cardinality $\mathfrak{c}$.*

COROLLARY (to Theorem 4 and Lemmas 2 and 2′).

1) [CH]. *$\beta N\backslash N$ is a weakly Noetherian compact Hausdorff space that is not Noetherian.*

2) [MA + ¬CH]. *$\beta N\backslash N$ is a compact Hausdorff space that is not weakly Noetherian.*

The assertion that $\beta N\backslash N$ is a weakly Noetherian compact Hausdorff space thus does not depend on the ZFC system of set theory axioms.

THEOREM 5. *If $\pi(w(X) \leq \aleph_0$, then $X\times(\beta N\backslash N)$ does not have a Noetherian π-base.*

PROOF Assume the opposite, and let $\mathcal{B}$ be a Noetherian π-base for $x \times (\beta N\backslash N)$. In each set in $\mathcal{B}$ we take a nonempty "rectangle" $P \times Q$, with P from some countable π-base σ for X and Q from some π-base δ for $\beta N\backslash N$. The family $\mathcal{D}$ of such "rectangles" is obviously Noetherian. Clearly, the family $\delta' = \{Q\colon P \times Q \in \mathcal{D}\}$ is a π-base for $\beta N\backslash N$, and thus, by Lemma 2, there exists a strictly decreasing subfamily δ'' that is of type ω_1 in δ'. Since $|\delta''| > \aleph_0$, there exist an element $P_0 \in \sigma$ and an uncountable subfamily δ''' of δ'' such that $P_0 \times Q \in \mathcal{D}$ for any $Q \in \delta'''$. This implies that $\mathcal{D}$ is not a weakly Noetherian family.

V. We investigate more closely spaces with weakly Noetherian π-bases.

THEOREM 6. *If $\pi\omega(X) < \aleph_1$, then X has a weakly Noetherian π-base.*

The proof presents no difficulties, and is omitted.

THEOREM 6′ [CH]. *A countable space has a weakly Noetherian π-base, as does a separable regular space.*

This follows easily from Theorem 6.

We prove below that under another hypothesis, namely, under [AM + ¬CH], a separable space does not necessarily have a weakly Noetherian π-base.

PROPOSITION 1. *Suppose that the topology τ is the supremum of a strictly increasing family A of topologies with countable π-bases. If $\xi \subset \tau$ and $\mathrm{cf}|A| > |\xi| \geq \mathrm{cf}|\xi| > \aleph_0$, then ξ contains a sequence ξ' of cardinality $|\xi|$ such that* $\mathrm{Int} \cap \xi' \neq \varnothing$.

For a proof it obviously suffices to confine ourself to the case when $\xi \subset \bigcup A$, but it is then clear that there exists a topology $\nu \in A$ such that $\xi \subseteq \nu$, and since ν has a countable π-base, there is an element of this π-base contained in each element of some subsystem ξ' of ξ with cardinality $|\xi|$ (because the π-base is countable, and $\mathrm{cf}|\xi| > \aleph_0$) which is what we were required to prove.

PROPOSITION 2. *Under the conditions of Proposition* 1 *suppose that* $\mathrm{cf}|A| > \aleph_1$ *and the topology* τ *does not have a countable* π*-base. Then the topology* τ *does not have a weakly Noetherian* π*-base.*

Indeed, if $\mathcal{P}$ is a π-base for τ, then, first, $|\mathcal{P}| > \aleph_0$, and, second, it can be assumed that $\mathcal{P} \subset \bigcup A$. Let ξ be a subsystem of $\mathcal{P}$ of cardinality $\aleph_1$; then all the conditions of Proposition 1 hold. Consequently, ξ contains a subsystem ξ' of uncountable cardinality such that $\mathrm{Int} \bigcap \xi' \neq \varnothing$, and this implies that $\mathcal{P}$ is not a weakly Noetherian family.

COROLLARY (to Propositions 1 and 2) [MA + $\neg$CH]. *On a countable set there exists a regular* T_1*-topology satisfying the conditions of Propositions* 1 *and* 2; *consequently, there exists a countable regular* T_1*-space and thus a separable compact Hausdorff space without a weakly Noetherian* τ*-base.*

Indeed, under MA (and even under a somewhat weaker assumption) we constructed in [9] a regular T_1-topology τ on a countable set X such that τ does not have a countable π-base but is the supremum of a strictly increasing well-ordered family of the type of the first ordinal with cardinality $\mathfrak{c}$ and consisting of topologies with countable π-bases. Under MA the cardinal $\mathfrak{c}$ is regular; therefore, for [MA + $\neg$CH] the topology τ satisfies all the conditions of Propositions 1 and 2. We remark that the space (X, τ) under discussion is also the space of a topological group.

The next theorem makes it easy for us to construct countable spaces without weakly Noetherian π-bases.

THEOREM 7 [MA]. *For every topology without isolated points and with a base of cardinality less than* $\mathfrak{c}$ *on a countable set there exists a larger* (*regular and* T_1-) *topology without isolated points and with a countable base.*

The proof is technically complicated and unconnected with the theme of this article, so is omitted.

Let us take a topology τ_0 without isolated points and with a countable base on a countable set. It is not hard to get a strictly increasing family $-\{\tau_\alpha\colon \alpha \in \omega_1\}$ of topologies with countable bases and without isolated points. the process of making the topologies stronger can proceed in such a way that the topology $\tau'_{\omega_1} = \bigcup\{\tau_\alpha\colon \alpha \in \omega_1\}$ does not have a countable τ-base. However, according to Theorem 7, there exists a topology τ_{ω_1} larger than it (and regular and T_1 if desired) with a countable base. The process of making the topologies stronger can now be continued further. If $\mathfrak{c} \geq \aleph_2$, then the topology $\bigcup\{\tau_\alpha\colon \alpha \in \omega_2\}$ is already the desired one: it does not have a weakly Noetherian π-base.

VI. We consider the following statement:

ASSERTION (*). *If a weakly Noetherian compact Hausdorff space X is irreducibly mapped onto a compact Hausdorff space Y of countable π-character satisfying the Suslin condition, then X and Y are compact metric spaces.*

Let φ be such an irreducible mapping. Fix some weakly Noetherian base $\mathcal{B}$ in X and let

$$\varphi\mathcal{B} = \{\operatorname{Int} \varphi B \colon B \in \mathcal{B}\}.$$

Obviously, the family $\varphi\mathcal{B}$ is weakly Noetherian, but since Y is of countable π-character, it is point-countable.

It is known (see [10]) that for [MA+¬CH] (and even under a somewhat weaker hypothesis) every point-countable family of open subsets in a complete (in the Čech sense) space with the Suslin condition is countable; therefore, for [MA + ¬CH] the family $\varphi\mathcal{B}$ is countable, and Assertion (*) is valid.

On the other hand, let SCH = the model for the ZFC system of set theory axioms in which there exists a Suslin continuum SC and CH is satisfied. Assertion (*) is not true in this model.

Indeed, it is shown below that the absolute of a Suslin continuum does not contain points of countable π-character. Therefore, according to Theorem 4, under CH the absolute $\overset{\circ}{\mathrm{SC}}$ of a Suslin continuum is a weakly Noetherian compact Hausdorff space; thus $\overset{\circ}{\mathrm{SC}}$, SC, and the natural mapping of $\overset{\circ}{\mathrm{SC}}$ onto SC satisfy the conditions of Assertion (*) under CH, but its conclusion is not satisfied.

Accordingly, we have proved that Assertion (*) does not depend on the ZFC system of set theory axioms.

It follows from the next proposition that $\overset{\circ}{\mathrm{SC}}$ does not have points of countable π-character.

PROPOSITION 3. *If a Hausdorff space X does not have separable open subsets, then in (the space of maximal systems of open subsets with the finite-intersection property) $\overset{\circ}{X}$ there are no points of countable π-character.*

PROOF. Let Δ be an arbitrary countable family of regular open subsets of X, and $A(\Delta)$ a countable subset of X obtained by choosing one point from each element of Δ. The set $A(\Delta)$ is countable and thus nowhere dense; consequently, for every point $\xi \in \overset{\circ}{X}$ there is a regular open set $B(\Delta, \xi)$ such that $B(\Delta, \xi) \in \xi$ and $B(\Delta, \xi) \cap A(\Delta) \varnothing$. This implies that

$$\operatorname{Int}(V \backslash B(\Delta, \xi)) \neq \varnothing$$

for any element $V \in \Delta$, i.e., the system $\{\overset{\circ}{V} \colon V \in \Delta\}$ does not form a π-base at any point of $\overset{\circ}{X}$. Proposition 3 is proved.

The author thanks S. A. Peregudov for useful discussions which facilitated the writing of this paper.

BIBLIOGRAPHY

1. A. V. Arkhangel'skiĭ and V. I. Ponomarev, *Foundations of general topology in problems and exercises*, "Nauka", Moscow, 1974; English transl., Hindustan, Delhi, and Reidel, Dordrecht, 1984.

2. S. A. Peregudov, *On certain properties of families of open sets and coverings*, Vestnik Moskov. Univ. Ser. I Mat. Mekh. **1976**, no. 3, 25–33; English transl. in Moscow Univ. Math. Bull. **31** (1976).

3. ——, *On* Π*-uniform bases and* π*-bases*, Dokl. Nauk SSSR **229** (1976), 542–545; English transl. in Soviet Math. Dokl. **17** (1976).

4. S. A. Peregudov and B. È. Shapirovskiĭ, *On a class of compact spaces*, Dokl. Akad. Nauk SSSR **230** (1976), 279–282; English transl. in Soviet Math. Dokl. **17** (1976).

5. R. M. Solovay and S. Tennenbaum, *Iterated Cohen extensions and Souslin's problem*, Ann. of Math. (2) **94** (1971), 201–245.

6. V. I. Malykhin and B. È. Shapirovskiĭ, *Martin's axiom and properties of topological spaces*, Dokl. Akad. Nauk SSSR **213** (1973), 532–535; English transl. in Soviet Math. Dokl. **14** (1973).

7. R. Engelking, *Outline of general topology*, PWN, Warsaw, Wiley, New York, and North-Holland, Amsterdam, 1968.

8. B. A. Efimov, *On extremally disconnected compact Hausdorff spaces*, Dokl. Akad. Nauk SSSR **172** (1967), 771–774; English transl. in Soviet Math. Dokl. **8** (1967).

9. V. I. Malykhin, *Extremally disconnected and similar groups*, Dokl. Akad. Nauk SSSR **220** (1975), 27–30; English transl. in Soviet Math. Dokl. **16** (1975).

10. B. È. Shapirovskiĭ, *On separability and metrizability of spaces with the Suslin condition*, Dokl. Akad. Nauk SSSR **207** (1972), 800–803; English transl. in Soviet Math. Dokl. **13** (1972).

Translated by H. H. McFADEN

Amer. Math. Soc. Transl.
(2) Vol. **134**, 1987

Cardinal Invariants in Compact Hausdorff Spaces

B. È. SHAPIROVSKIĬ

Here we study cardinal characteristics of topological spaces such as the density, Souslin number, tightness, character, π-weight ($s(X)$, $c(X)$, $t(X)$, $\chi(X)$, $\pi w(X)$, respectively), and others, as well as their hereditary analogues.

In particular, the following result is proved in §1.

COROLLARY 1.1. *A compact Hausdorff space X with countable tightness is separable if (and only if) the cardinal $\aleph_1$ is a caliber for X.*

This corollary, which improves a number of known results of Arkhangel'skiĭ [7], Juhász [60], and Tall [77], has no forerunner even in the class of compact Hausdorff spaces of countable character. Therefore, we especially single out

COROLLARY 1.1′. *A compact Hausdorff space with the first axiom of countability such that $\aleph_1$ is a caliber is separable.*

A problem of Hajnal and Juhász stated in the lecture notes of M. E. Rudin [72] is solved in §2. Namely,

THEOREM 2.1. *If X is a compact Hausdorff space, then $\bar{s}(X) \le \bar{c}(X)^+$.*

An essential refinement of Theorem 2.1 is given by

THEOREM 2.1°. *If X is a compact Hausdorff space, then $\bar{s}(X) \le t(X)^-\bar{c}(X)$.*

Indeed, Theorem 2.1 makes possible a number of unexpected conclusions:

COROLLARY 2.3°. *If a compact Hausdorff space with countable tightness contains an uncountable discrete subset, then its hereditary Suslin number and its hereditary density coincide.*

1980 *Mathematics Subject Classification* (1985 *Revision*). Primary 54A25, 54D30; Secondary 54D15, 54D20, 54D55, 54E65.

Translation of Seminar on General Topology (P. S. Aleksandrov, editor), Moskov. Gos. Univ., Moscow, 1981, pp. 162–187; MR **83f**:54024.

THEOREM 2.5. *The hereditary Suslin number for a compact Hausdorff space is always equal to its hereditary density or tightness.*

In the compact case these results practically answer a question of de Groot [49], since (as follows from [56] and [78]) it is not possible to prove the equality $\bar{s}(X) = \bar{c}(X)$ by naive means even in the class of compact Hausdorff spaces([1])

Finally, in §3 we prove

THEOREM 3.1. *In each compact Hausdorff space X there exists a π-base whose multiplicity does not exceed $t(X)$.*

This theorem gives the key simultaneously to the proofs of both the results in §1 and those in §2. Indeed, by the definition of a caliber, Corollary 1.1 follows immediately from Theorem 3.1. On the other hand, the following result is also based on Theorem 3.1.

THEOREM 3.2. *If X is a compact Hausdorff space, then $\overline{ds}(X) \leq t(X)^+$ and, hence, $\bar{s}(X) \leq t(X)^+ \bar{c}(X)$.*

Here $ds(X) = \min\{\tau\colon$ there exists a family $\mathcal{D}$ of discrete subsets of X such that $X = \operatorname{Cl} \bigcup \mathcal{D}$, $|D| \leq \tau\}$ is the d-density of X (see, [12] and [28]).

Thus, it is possible to establish a relation between a local invariant—the tightness (character)—on one hand, and a global invariant—the d-density—on the other.

ASSERTION 1°. *If $A' = \operatorname{Cl}_\tau A$ and $s(A') \leq \tau$, then $s(A) \leq \tau$.*([2])

Indeed, suppose that $S \subset A'$ and $\operatorname{Cl} S \supset A'$. Then for every point $x \in S \subset A'$ there exists a $C_x \subset A$ such that $x \in \operatorname{Cl} C_x$ and $|C_x| \leq \tau$. It is clear that, setting $C = \bigcup\{C_x\colon x \in S\}$, we get $|C| \leq |S| \cdot \tau$, $\operatorname{Cl} C \supset \operatorname{Cl} S \supset A' \supset A$, and $C \subset A$, which implies the required inequality when $|S| \leq \tau$.

Since $\operatorname{Cl}_{t(X)} A = \operatorname{Cl} A$, an immediate consequence of Assertion 1° is

ASSERTION 1. *$s(A) \leq s(\operatorname{Cl} A) \cdot t(X)$, and, hence, $\bar{s}(X) = s'(X) \cdot t(X)$ (cf. [6], [32], and [60]).*

We need the following result, which was proved in [31] and independently in [6] (in this connection see [39], [40], and [36]).

R1. *If X is a compact Hausdorff space, then $t(X) \leq \bar{c}(X)$*

An immediate consequence of Assertion 1 and R1 is

R2. *If X is a compact Hausdorff space, then $s'(X) = \bar{s}(X)$ ([6], [32], [36]).*

Moreover, we need the following two results.

R3. *If X is a compact Hausdorff space, then $t(X) = \overline{\pi\chi}(X)$ [37].*

R4. *If X is a compact Hausdorff space, then $\bar{s}(X) = \overline{\pi w}(X)$ [37].*

([1])For a cardinal-valued function $\tau(Y)$, $Y \subset X$, we set

$$\bar{\tau}(X) = \sup\{\tau(Y)\colon Y \subset X\} \quad \text{and} \quad \tau'(X) = \sup\{\tau(F)\colon \operatorname{Cl} F = F \subset X\}.$$

Thus, the quantities $s'(X)$, $\bar{s}(X)$, $\bar{c}(X)$, $\bar{\pi}\omega(X)$, etc., are defined.

([2])$\operatorname{Cl}_\tau A = \bigcup\{\operatorname{Cl} M\colon M \subset A,\ |M| \leq \tau\}$.

We also state the simple

ASSERTION 2°. *If* $\{A_\alpha\colon \alpha < \tau\}$ *is a nondecreasing family of subsets of* X, $A = \bigcup\{A_\alpha\colon \alpha < \tau\}$, *and* $\nu < \mathrm{cf}(\tau)$, *then* $\mathrm{Cl}_\nu A = \bigcup\{\mathrm{Cl}_\nu A_\alpha\colon \alpha < \tau\}$.

Indeed, if $x \in \mathrm{Cl}_\nu A$, then there exists a set $C_x \subset A$ such that $x \in \mathrm{Cl}\, C_x$ and $|C_x| \le \nu$. But since $\nu < \mathrm{cf}(\tau)$, it follows that $C_x \subset A_{\bar\alpha}$ for some $\bar\alpha < \tau$; hence, $x \in \mathrm{Cl}\, C_x \subset \mathrm{Cl}_\nu A_{\bar\alpha}$, which implies the required equality.

Setting $\nu = t(X)$ in Assertion 2°, we immediately get

ASSERTION 2. *If* $\{F_\alpha\colon \alpha < \tau\}$ *is a nondecreasing family of closed subsets of* X *and* $t(X) < \mathrm{cf}(\tau)$, *then the set* $F = \bigcup\{F_\alpha\colon \alpha < \tau\}$ *is closed in* X.

For a family $\mathcal{A}$ of subsets of X we set

$$\bigvee_\tau \mathcal{A} = \left\{\bigcup \mathcal{A}'\colon \mathcal{A}' \subset \mathcal{A},\ |\mathcal{A}'| \le \tau\right\},$$

$$\bigwedge_\tau \mathcal{A} = \left\{\bigcap \mathcal{A}'\colon \mathcal{A}' \subset \mathcal{A},\ |\mathcal{A}'| \le \tau\right\},$$

$$\bigvee \mathcal{A} = \bigcup\left\{\bigvee_n \mathcal{A}\colon n < \omega_0\right\} = \left\{\bigcup \mathcal{A}'\colon \mathcal{A}' \subset \mathcal{A},\ |\mathcal{A}'| < \aleph_0\right\},$$

$$\bigwedge \mathcal{A} = \bigcup\left\{\bigwedge_n \mathcal{A}\colon n < \omega_0\right\} = \left\{\bigcap \mathcal{A}'\colon \mathcal{A}' \subset \mathcal{A},\ |\mathcal{A}'| < \aleph_0\right\}.$$

Clearly, $\bigwedge \mathcal{A}$ (respectively, $\bigvee \mathcal{A}$) is the family of all subsets of X which are intersections (unions) of finite subfamilies of $\mathcal{A}$.

For brevity the following notation is also used:

$$\mathrm{Cl}\mathcal{A} = \{\mathrm{Cl}\, A\colon A \in \mathcal{A}\}, \qquad \mathcal{A}_x = \{A \in \mathcal{A}\colon A \in X\}.$$

A family $\mathcal{A}$ is said to be point-countable (point-finite) if $|\mathcal{A}_x| \le \aleph_0$ (respectively, $|\mathcal{A}_x| < \aleph_0$) for all $x \in X$.

A regular cardinal τ is said to be a *caliber* [30] for a space X if every family $\mathcal{J}$ of cardinality τ consisting of open subsets of X(3) contains a subfamily $\mathcal{J}' \subset \mathcal{J}$ of the same cardinality τ and with nonempty intersection.

A cardinal τ is called a *precaliber* for X if every family $\mathcal{I}$ of cardinality τ consisting of open subsets of X contains a subfamily $\mathcal{I}' \subset \mathcal{I}$ of the same cardinality τ with the finite-intersection property (see [77]). Let

$$\mathrm{Sh}(X) = \sup\{\tau\colon \tau \text{ is regular and is not a caliber for } X\},$$

the *Shanin number* [32].

If $\mathrm{Sh}(X) \le \aleph_0$, then X is said to satisfy the *Shanin condition* [9]. Clearly, $\mathrm{Sh}(X) = \min\{\tau\colon$ every regular cardinal $\nu > \tau$ is a caliber for $X\}$. Thus, a space X satisfies the Shanin condition if every uncountable regular cardinal is a caliber for X. The Shanin condition is known to be satisfied by all separable spaces, as well as all the spaces obtained by taking the closure of the class of separable spaces with respect to the operations of taking products (with any

(3)Families of open sets will always be assumed not to contain empty sets.

number of factors), taking continuous images, and passing to open subsets. All dyadic compact Hausdorff spaces, for example, satisfy the Shanin condition.

However, in this article a weaker condition will suffice for us. Let

$$\mathrm{Sh}_0(X) = \min\{\tau\colon \text{for all } \nu \geq \tau \text{ the cardinal } \nu^+ \text{ is a caliber for } X\},$$

the *shaliber* of a space X.

Clearly, $\mathrm{Sh}_0(X) \leq \mathrm{Sh}(X)$ and, consequently, every space with the Shanin condition is obviously a space of countable caliber.

We formulate a number of simple assertions connected with these concepts.

ASSERTION 3. *A regular cardinal τ is a caliber for X if and only if $|\mathcal{B}| < \tau$ for every family $\mathcal{B}$ of open subsets of X such that $|\mathcal{B}_x| < \tau$ for any $x \in x$.*

An immediate consequence of Assertion 3 is

ASSERTION 3′. *If $\mathcal{D}$ is a point-countable (disjoint) family of open subsets of X and τ is caliber (precaliber) for the Hausdorff space X, then $|\mathcal{D}| < \tau$.*

On the other hand, if $\mathrm{Cl}\, S = X$ and the regular cardinal τ satisfies $\tau > |S|$, then obviously τ is a caliber for X, and by considering Assertion 3′ we easily get

ASSERTION 4. $c(X) \leq \min\{\tau\colon \tau \text{ is a precaliber for } X\} \leq \mathrm{Sh}_0(X) \leq \mathrm{Sh}(X) \leq s(X)$.

A nondecreasing family $\mathcal{F} = \{F_\alpha\colon \alpha < \tau\}$ of closed subsets of X is called a *tower of height* τ (or a τ*-tower*) in X if $\bigcup \mathcal{F} = X$ and $F_\alpha \neq X$ for all $\alpha < \tau$.

ASSERTION 5. *A regular cardinal τ is not a caliber if and only if there exists a τ-tower in X.*

Indeed, if $\mathcal{F}$ is a τ-tower in X, then, setting $\mathcal{B} = \{X \backslash F_\alpha\colon F_\alpha \in \mathcal{F}\}$, we get that $|\mathcal{B}| = \tau$, but $|\mathcal{B}_x| < \tau$ for all $x \in X$.

ASSERTION 6. *If X is a compact Hausdorff space and the regular cardinal τ is a precaliber for X, then τ is a caliber for X (see [60]).*

Indeed, if $\mathcal{J}$ is a family of open subsets of X and $|\mathcal{J}| = \tau$, then, forming the family $\mathcal{V} = \{V_U\colon U \in \mathcal{J}\}$, where $\varnothing \neq V_U \subset \mathrm{Cl}\, V_U \subset U$, and singling out a subfamily $\mathcal{V}' \subset \mathcal{V}$ with the finite-intersection property and such that $|\mathcal{V}| = |\mathcal{V}'| = \tau$, we get the desired family $\mathcal{J}' = \{U\colon V_U \in \mathcal{V}'\}$, because the compactness of X implies that $\varnothing \neq \bigcap[\mathcal{V}'] \subset \bigcap \mathcal{J}'$, and $|\mathcal{J}'| = |\mathcal{V}'| = \tau$.

§1. Calibers and density in compact Hausdorff spaces. Martin's axiom

The central result of this section (Theorem 1.1) has as corollaries several well-known results of Arkhangel′skiĭ [7], [6], Juhász [60], [59], and Tall [77].

THEOREM 1.1. *Suppose that X is a compact Hausdorff space, the regular cardinal τ is a (pre)caliber for X, and $t(X) < \tau$. Then $s(X) < \tau$ and, consequently, $s(X) \leq t(X)$ if $t(X)^+$ is a (pre)caliber for X.*

PROOF. Assuming the opposite, namely, $s(X) \geq \tau$, we carry out a transfinite construction. Take an arbitrary point $\bar{x} \in X$ and a nonempty open subset U of X such that $\bar{x} \notin \operatorname{Cl} U$, let $F_0 = \{\bar{x}\}$ and $U_0 = U$, and assume that for every $\alpha < \alpha' < \tau$ we have already determined a compact set $F_\alpha \subset X$ and a nonempty open set $U_\alpha \subset X$ such that:

(1) $s(F_\alpha) < \tau$;

(2) $F_\alpha \cap \operatorname{Cl} U_\alpha = \varnothing$ and $F_\alpha \subset F_\beta$ for $\alpha < \beta < \tau$; and

(3) if $\bigcap \mathcal{K} \neq \varnothing$ for a finite family $\mathcal{K} = \{U_{\alpha_1}, \ldots, U_{\alpha_k}\}$, and $\max\{\alpha_1, \ldots, \alpha_k\} + 1 = \bar{\alpha} < \alpha'$, then $F_{\bar{\alpha}} \cap \bigcap \mathcal{K} \neq \varnothing$.

We construct the next sets $F_{\alpha'}$, and $U_{\alpha'}$. Let $\mathcal{J}_{\alpha'} = \{U_\alpha \colon \alpha < \alpha'\}$ and, fixing a point $x(G) \in G$ for every nonempty set $G \in \bigwedge \mathcal{J}_{\alpha'}$, form the set

$$S_{\alpha'} = \{x(G) \colon G \in \bigwedge \mathcal{J}_{\alpha'},\ G \neq \varnothing\}.$$

It is clear that $|S_{\alpha'}| \leq |\bigwedge \mathcal{J}_{\alpha'}| = |\mathcal{J}_{\alpha'}| = |\alpha'| < \tau$ ($\alpha' < \tau$, and hence $|\alpha'| < \tau$). Moreover, τ is a regular cardinal, and, consequently, by (1),

$$\sum \{s(F_\alpha) \colon \alpha < \alpha'\} < \tau$$

Therefore, setting $F_{\alpha'} = \operatorname{Cl}(\bigcup\{F_\alpha \colon \alpha < \alpha'\}) \cup \operatorname{Cl} S_{\alpha'}$, we get that $s(F_{\alpha'}) < \tau$. But by assumption $s(X) \geq \tau$; hence, $X \backslash F_{\alpha'} \neq \varnothing$, and there exists a nonempty open subset V of X for which $F_{\alpha'} \cap \operatorname{Cl} V = \varnothing$. Let $U_{\alpha'} = V$. Obviously, $F_{\alpha'}$ and $U_{\alpha'}$ are the desired sets, because condition (3) is also satisfied, by the choice of the set $S_{\alpha'}$.

We can thus regard the family $\mathcal{F} = \{F_\alpha \colon \alpha < \tau\}$ of compact sets and the family $\mathcal{I} = \{U_\alpha \colon \alpha < \tau\}$ of open sets corresponding to conditions (1)–(3) as constructed.

Since τ is a precaliber for X, there exists a family $\mathcal{I}' \subset \mathcal{I}$ with the following properties.

(a) $\mathcal{I}'$ is a family with the finite-intersection property.

(b) $|\mathcal{I}'| = \tau$.

We show that the existence of such a family leads to a contradiction. Indeed, by Assertion 2, the set $\bigcup\{F_\alpha \colon \alpha < \tau\} = F$ is closed in X, because $t(X) < \tau = \operatorname{cf}(\tau)$ and $\mathcal{F}$ is an increasing family. Therefore, it follows from the fact that X is compact and the conditions (a) and (3) that $F \cap \bigcap \operatorname{Cl} \mathcal{I}' \neq \varnothing$. But, on the other hand, by (b), for every $\alpha < \tau$ there exists a $U_\beta \in \mathcal{I}'$ such that $\beta > \alpha$ and, consequently, by (2), $F_\alpha \cap \operatorname{Cl} U_\beta = \varnothing$; therefore, $F \cap \bigcap \operatorname{Cl} \mathcal{I}' = \varnothing$, a contradiction. Accordingly, $s(X) < \tau$.

COROLLARY 1.1. *If X is a compact Hausdorff space with countable tightness and $\aleph_1$ is a caliber for X, then X is separable.*

We single out as a special case

COROLLARY 1.1′. *A compact Hausdorff space with the first axiom of countability such that $\aleph_1$ is a caliber is separable.*

Our result is new even for this special case.

Kunen and, independently, Solovay and Rowbottom (see [59]) established the following important fact, which is directly related to our topic.

K_{ω_1}. Under the premise MA&($\aleph_1 < \mathfrak{c}$), $\aleph_1$ is a precaliber for every space with the Suslin condition(4) (i.e., every uncountable family of open subsets of X contains an uncountable subfamily with the finite-intersection property).

As already mentioned, in the compact case every regular precaliber is obviously a caliber, and thus K_{ω_1} implies $\overline{K}_{\omega_1}$: Under the premise MA&($\aleph_1 < \mathfrak{c}$), $\aleph_1$ is a caliber for every compact Hausdorff space with the Suslin condition (see [60], [53], and also [6] and [77]).(5)

From K_{ω_1} and Corollary 1.1 we now immediately get

COROLLARY 1.2. *Under the premise MA&($\aleph_1 < \mathfrak{c}$) every compact Hausdorff space with the Suslin condition and countable tightness is separable.*

For compact Hausdorff spaces with the first axiom of countability an analogous result, also based on the assertion K_{ω_1}, was obtained by Juhász [60]. However, this proof, unlike that of Theorem 1.1, uses the fact that not only a compact Hausdorff space X with the Suslin condition, but also any closed subset of it (of density $\aleph_1$) also having the Suslin condition has $\aleph_1$ as a caliber. (Juhász's proof also uses the first axiom of countability in full measure in order to construct a compact set $F \subset X$ with $c(F) \leq \aleph_0$ and $s(F) = \aleph_1$.) Juhász's theorem was improved by Arkhangel′skiĭ [7], who proved (also with the aid of the Kunen-Rowbottom-Solovay result) an analogous assertion for sequential compact Hausdorff spaces, though under a somewhat more stringent axiomatic premise. In a more general formulation the theorem of Arkhangel′skiĭ is as follows ([7], Theorem 2):

If X is a compact Hausdorff space of countable tightness, $c(X) \leq \aleph_0$, and $|X| \leq \mathfrak{c}$, then X is separable under the premise MA&($\aleph_2 = \mathfrak{c}$).

Arkhangel′skiĭ also posed a problem [39] which is solved positively by Corollary 1.2: the restrictions on the cardinality in Theorem 2 of [7] can actually be gotten rid of (and instead of "$\aleph_2 = \mathfrak{c}$" it suffices to require "$\aleph_2 \leq \mathfrak{c}$"). We remark also that the question of the possibility of replacing the character by the tightness in Juhász's theorem figures in the thorough treatment of Tall [77]. Further, we note that, by Corollary 1.2 and R1 ($t(X) \leq \bar{c}(X)$), for $c(X) \leq \aleph_0$ each closed subspace (and hence, by R2, each subspace) $A \subset X$ is separable. Thus, from Corollary 1.2, R1, and R2 we immediately get another theorem of Arkhangel′skiĭ [6].

(4)In recent years there have appeared many articles about Martin's axiom (MA), which has found application in diverse areas of general topology (see, for example, [60], [72], [7], [6], [53], [59], [41], [77], [21], and [34]).

(5)The fact that Martin's axiom combined with the negation of the continuum hypothesis (MA&($\aleph_1 < \mathfrak{c}$)) is independent of the commonly used ZFC system of set theory axioms was proved in [64] and [73]. For details see [19] or, for example, [72]. The formulation of MA can be found in, for example, [19] or [72], [60], or [21].

Under the premise MA&($\aleph_1 < \mathfrak{c}$) every discrete-countable(6) *compact Hausdorff space is hereditarily separable.*

An analogous result was obtained by Juhász [59] for perfectly normal compact Hausdorff spaces. Together with Hajnal he proved the following [53]:

Under MA a compact Hausdorff space X with the Suslin condition and such that $\pi w(X) < \mathfrak{c}$ is separable.

On the basis of this useful result, Tall proved in [77] (Theorem 4.14, Corollary 4.15) that:

Under MA a compact Hausdorff space X with the Suslin condition such that $s(X){\cdot}t(M) < \mathfrak{c}$ and $|X| < 2^{\mathfrak{c}}$ is separable; in particular, X is separable if $\bar{s}(X) < \mathfrak{c}$ and $|X| < 2^{\mathfrak{c}}$.

It turns out, however, that we can get rid of the restriction on the cardinality ($|X| < 2^{\mathfrak{c}}$) also in this result. Indeed, by R3, $\pi w(X) = s(X){\cdot}\pi\chi(X) \leq s(X){\cdot}t(X)$, and with the help of the cited theorem of Hajnal and Juhász in [53] we get

ASSERTION 1.1. *Under MA a compact Hausdorff space with the Suslin condition and such that $s(X){\cdot}t(X) < \mathfrak{c}$ is separable, and if $\bar{s}(X) < \mathfrak{c}$, then $s(X) \leq \aleph_0$.*

COROLLARY 1.3. *If X is a compact Hausdorff space with the Suslin condition and countable tightness, then "$s(X) < \mathfrak{c}$" implies "$s(X) \leq \aleph_0$" under MA.*

Formally weaker than Corollary 1.2, Corollary 1.3 (unlike the former) does not assume the negation of the continuum hypothesis.

REMARK. In [21] (Theorem 3.1 and Corollary 3.1 in §3) we improved the Kunen-Rowbottom-Solovay result (K_{ω_1}), namely, we proved that if $c(X) \leq \aleph_0$, then under MA every family $\mathcal{B}$ of open subsets of X such that $|\mathcal{B}| = \tau < \mathfrak{c}$ decomposes into a countable system $\{\mathcal{B}_i : i < \omega_0\}$ of families with the finite-intersection property; consequently, if $\aleph_0 < \mathrm{cf}(\tau) \leq \tau < \mathfrak{c}$, then τ is a precaliber for X. This result, in turn, is a consequence of the main Lemma 3.1 in [21] (§3), which enables us, moreover, to carry all the aforementioned results proved under MA or MA&($\aleph_1 < \mathfrak{c}$) (for example, the theorem of Hajnal and Juhász) over to broader classes of spaces, and, in particular, to spaces complete in the Čech sense (see [34], [21], and [77]). (We remark that the result of Tall in [77] was also formulated for G_λ-spaces with $\lambda < \mathfrak{c}$.) However, we do not give these facts our attention, since a more detailed discussion of questions of this kind does not enter into our problem.

It should be mentioned that all the results proved above under MA or MA&($\aleph_1 < \mathfrak{c}$) are not only consistent, but also do not depend on the common set theory axioms: Jech [56] and Tennenbaum [78] have constructed models in which there exists a Suslin continuum(7) (i.e., a totally ordered connected nonseparable compact Hausdorff space with the Suslin condition). What is more,

(6)A space X is discrete-countable if $\bar{c}(X) \leq \aleph_0$. Such spaces are also called hereditarily Suslin spaces (see, for example, [31]–[36]).

(7)A remarkable result in this direction was obtained by Jense [57]: the constructivity axiom implies the existence of a Suslin continuum.

the continuum hypothesis (CH) is satisfied in these models, and, consequently (since CH obviously implies MA; see, for example, [19] or [72]), it is necessary to assume the negation of the continuum hypothesis ($\aleph_1 < \mathfrak{c}$) to prove Corollary 1.2. On the other hand, as proved by Solovay and Tennenbaum [73], under the premise MA&($\aleph_1 < \mathfrak{c}$) every totally ordered space with the Suslin condition is separable. Thus, the problem posed as far back as 1920 by Suslin [74], and about which more than enough has been written in the course of a half-century (see, for example, [71] and [19]), receives its final solution. We remark that every totally ordered space with the Suslin condition has a totally ordered perfectly normal (and hence with $t(X) \leq \chi(X) \leq \aleph_0$) compactification (see [71] and [52]), and, as shown above, Corollary 1.2 is the strongest result concerned with conditions under which $c(X) \leq \aleph_0$ implies separability.

PROPOSITION 1.1. *Suppose that X is a compact Hausdorff space and $t(X) \leq \tau$. Then the following conditions are equivalent:*

(a) *If the multiplicity of a family $\mathcal{B}$ of open subsets of X is at most τ, then $|\mathcal{B}| \leq \tau$.*

(b°) $\mathrm{Sh}_0(X) \leq \tau$.

(b) $\mathrm{Sh}(X) \leq \tau$.

(c) $s(X) \leq \tau$.

(d) $\pi w(X) \leq \tau$.

(e) *Every family $\mathcal{B}$ of cardinality $\geq \tau^+$ consisting of open subsets of X contains a subfamily $\mathcal{B}'$ of cardinality $\geq \tau^+$ such that $\mathrm{Int}(\bigcap \mathcal{B}') \neq \varnothing$.*

(f) *Every family $\mathcal{B}$ of cardinality $\geq \tau^+$ consisting of open subsets of X contains a subfamily $\mathcal{B}'$ of cardinality $\geq \tau^+$ such that $\bigcap \mathcal{B}' \neq \varnothing$ (i.e., τ^+ is a caliber).*

(g) *Every family $\mathcal{B}$ of cardinality $\geq \tau^+$ consisting of open subsets of X contains a subfamily $\mathcal{B}'$ of cardinality $\geq \tau^+$ with the finite-intersection property (i.e., τ^+ is a precaliber).*

PROOF. It is clear that (d)$\Longrightarrow$(c)$\Longrightarrow$(b)$\Longrightarrow$(b°), and (b°)$\Longrightarrow$(a), because, by Assertion 3, (a) means that τ^+ is a caliber for X. On the other hand, (a)$\Longrightarrow$(c) by Theorem 1.1, and, since (by R3) $\pi w(X) \leq s(X) \leq \tau$, we have (a)$\Longrightarrow$(d). Further, it is obvious that (e)$\Longrightarrow$(f)$\Longrightarrow$(g) and (g)$\Longrightarrow$(a), by Assertion 6 and the fact that X is a compact Hausdorff space. It thus remains to show that (d)$\Longrightarrow$(e). But, if $\mathcal{I}$ is a π-base in X and $|\mathcal{I}| \leq \tau$, then, setting $\mathcal{B}_U = \{B \in \mathcal{B} : B \supset U\}$, we obviously get that $\mathcal{B} = \bigcup\{\mathcal{B}_U : U \in \mathcal{I}\}$. Since $|\mathcal{I}| \leq \tau < \tau^+ \leq |\mathcal{B}|$, there is a nonempty set $U' \in \mathcal{I}$ such that $|\mathcal{B}_{U'}| \geq \tau^+$, and, since $\mathrm{Int}(\bigcap \mathcal{B}_{U'}) \supset U' \neq \varnothing$, $\mathcal{B}_{U'}$ is the desired family.

COROLLARY 1.4. *If X is a compact Hausdorff space with countable tightness and every uncountable family of open subsets of X contains an uncountable subfamily with the finite-intersection property, then every uncountable family of open subsets of X contains an uncountable subfamily whose intersection has nonempty interior.*

We remark that, in view of the equivalence of (c) and (d) in Proposition 1.1, not only separability but also countable π-weight is proved in each of Corollaries 1.1–1.3.

Setting $\tau = \mathrm{Sh}_0(X) \cdot t(X)$ in Proposition 1.1, we immediately get

THEOREM 1.2. *Suppose that X is a compact Hausdorff space. Then* $\pi w(X) \le \mathrm{Sh}_0(X) \cdot t(X)$.

COROLLARY 1.5. *If X is a compact Hausdorff space with the Shanin condition, then* $\pi w(X) \le t(X)$.

By the obvious inequality $\mathrm{Sh}(X) \cdot t(X) \le w(X)$, Theorem 1.2 also implies

COROLLARY 1.6. *If the π-weight of a compact Hausdorff space X is equal to its weight, then* $w(X) = \mathrm{Sh}_0(X) \cdot t(X)$.

We say that a Hausdorff space X is an Sh *π-space* if (a) $\mathrm{Sh}_0(X) \le \aleph_0$ and (b) $\pi w(X) = w(X)$.

It is clear that Corollary 1.6 combined with R1 now implies

COROLLARY 1.7. *If X is a compact* Sh *π-space, then* $w(X) = t(X)$, *and, all the more so,* $w(X) = \bar{c}(X)$.

We have singled out the class of Sh π-spaces in order to demonstrate the large margin by which Corollary 1.7 (which is actually a simple deduction from Theorem 1.1) solves a problem posed by Arkhangel′skiĭ and Ponomarev in 1968: does the equality $w(X) = t(X)$ hold for every compact Hausdorff space that is simultaneously a κ-space and a π-space ([9], Question 1)? Indeed, the concept of an Sh π-space is an essential weakening of the concept of a π-space in the sense of Arkhangel′skiĭ and Ponomarev [9], and the condition of being a κ-space turns out to be superfluous in general.

The equality $w(X) = \pi w(X)$ [22] is also satisfied for dyadic compact Hausdorff spaces, which, as has been noted, satisfy the Shanin condition [30]; in this connection see also [45]. Moreover, Dantov* spaces in the Arkhangel′skiĭ sense [3] are also Sh π-spaces. Indeed, it is easy to prove that $\pi w(X) = w(X)$ for every thick (and hence, for every Dantov) space X ([3], Corollary 1 to Theorem 7).

Moreover, we have the simple

ASSERTION 1.2. *If X $\aleph_0$-suppresses (in the sense of* [3], §2) *a set $S \subset X$ and* $\mathrm{Cl}\, S = X$, *then* $\mathrm{Sh}_0(X) \le \aleph_0$.

PROOF. Suppose that $\mathcal{B}$ is a family of open subsets of X and $|\mathcal{B}| = \lambda^+ \ge \aleph_1$. For $B \in \mathcal{B}$ we fix a point $x(B) \in B \cap S$ and let $A = \{x(B) \colon B \in \mathcal{B}\}$. Since $A \subset S$ and $|A| \le |\mathcal{B}| \le 2^\lambda$, there exists by definition ([3], §2) an $A' \subset X$ such that $|A'| \le \lambda$ and $\mathrm{Cl}\, A' \supset A$. It is clear that there then exists a point $x' \in A'$ such that $|\mathcal{B}_{x'}| = |\mathcal{B}| = \lambda^+$ and, hence, λ^+ is a caliber for X.

* *Editor's note.* The Russian is "дантово", which is the neuter adjectival form of a proper name, probably not "Dantov". In a literary context "дантоз" is a Russian rendering of "Dante."

Thus, from very general arguments (Theorem 1.1 and Corollary 1.7) we easily deduce the following assertion:

If X is a Dantov space, then $w(X) = t(X)$ [3] and, a fortiori $w(X) = \bar{c}(X)$; in particular, if X is a dyadic compact Hausdorff space, then $w(X) = t(X)$ ([9], [3]) and, a fortiori, $w(X) = \bar{c}(X)$ [15].

To conclude this section we give one more result which immediately follows from Assertion 4, Proposition 1.1, and R1.

THEOREM 1.3. *Suppose that* τ^+ *is a caliber for every closed subset of a compact Hausdorff space* X. *Then* $\mathrm{Sh}'_0(X) = s'(X) = \overline{\pi w}(X) \le \tau$.

This theorem can be proved by using only the assertions from the introduction and without appealing to Proposition 1.1 (which follows from Theorem 1.1).

Indeed, if there exists a set $A' \subset X$ such that $\pi w(A') > \tau$, then by R4 there exists a set $A \subset X$ such that $s(A) > \tau$. But then it is clear that there exists a set $S = \{x_\alpha : \alpha < \tau^+\}$ such that $x_\beta \neq F_\beta = \{x_\alpha : \alpha < \beta\}$ for all $\beta < \tau^+$. By R1 and Assertion 4, $t(X) \le \bar{c}(X) = c'(X) \le \tau$. Therefore, $\mathrm{Cl}\, S = \bigcup\{F_\beta : \beta < \tau^+\}$ by Assertion 2, and, since the family $\{F_\beta : \beta < \tau^+\}$ is easily seen to be a τ^+-tower in $\mathrm{Cl}\, S$, the cardinal τ^+ is not a caliber for $\mathrm{Cl}\, S$ in view of Assertion 5. This is a contradiction. Consequently, $\overline{\pi w}(X) \le \tau$. Further, the inequalities $\mathrm{Sh}'(X) \le s'(X) \le \overline{\pi w}(X)$ are obvious. On the other hand, $\mathrm{Sh}'_0(X)^+$ is by definition a caliber for every closed $F \subset X$, and hence $\overline{\pi w}(X) \le \mathrm{Sh}'_0(X)$ by what was proved above. Accordingly, $\mathrm{Sh}'_0(X) = \overline{\pi w}(X) \le \tau$.

COROLLARY 1.8. *If* X *is a compact Hausdorff space, then* $\overline{\pi w}(X) \le \aleph_0$ *if and only if* $\aleph_1$ *is a (pre) caliber for every closed subset of* X.

§2. An upper bound for cardinalities of discrete subspaces, and density. Pseudocharacter of closed subsets and the cardinality of compact Hausdorff spaces

The *hereditary cellularity* (or *hereditary Suslin number*) $\bar{c}(X)$ of a space X is defined as $\bar{c}(X) = \sup\{c(Y) : Y \subset X\}$. On the other hand, it is perfectly obvious that $\bar{c}(X) = \sup\{|D| : D \text{ is a discrete subset of } X\}$, the supremum of the cardinalities of the discrete subsets of X.

The relation established in this section between the hereditary cellularity $\bar{c}(X)$ and the hereditary density $\bar{s}(X)$ (Theorem 2.3) answers positively a question of Hajnal and Juhász stated in M. E. Rudin's lecture notes [72]. This result implies a fact that is new even in the class of perfectly normal compact Hausdorff spaces:

The density (and even the hereditary density) of a perfectly normal compact Hausdorff space does not exceed $\aleph_1$ (*Corollary* 2.1).

In passing we use the same method to prove the well-known theorem of Arkhangel′skiĭ [4] on the cardinality of a compact Hausdorff space with the first axiom of countability.

A family $\mathcal{B}$ of open subsets of X is called a *pseudobase for a set* $A \subset X$ in X if for every point $x \in X \backslash A$ there is a set $B \in \mathcal{B}$ such that $A \subset B \subset X \backslash \{x\}$

[1]. Thus, a family $\mathcal{B}$ is a pseudobase for A in X if and only if there exists a subfamily $\mathcal{B}' \subset \mathcal{B}$ such that $\bigcap \mathcal{B}' = A$.

If $\mathcal{B}$ is a pseudobase for each point $x \in X$ in X, then we call $\mathcal{B}$ a *pseudobase for* X (or in X).

$\psi(A, X) = \min\{|\mathcal{B}|: \mathcal{B}$ is a pseudobase for A in $X\}$ is the *pseudocharacter* of a set A in X.

$\psi(X) = \sup\{\psi(x, X): x \in X\}$ is the *pseudocharacter of* X.(8)

THEOREM 2.1. *Suppose that X is a compact T_1-space and $\psi(F, X) \leq \mu$ for every closed $F \subset X$ such that $s(F) \leq \mu$. If $t(X)^+ = \nu \leq \mu$, then $\bar{s}(X) \leq \mu$.*

PROOF. We carry out a transfinite induction. Take an arbitrary point $x \in X$ and a pseudobase $\mathcal{B}_x$ for x in X such that $|\mathcal{B}_x| \leq \mu$, and let $F_0 = \{x\}$ and $\mathcal{B}_0 = \mathcal{B}_x$. Suppose that for all $\alpha < \alpha' < \nu$ we have already determined a closed set $F_\alpha \subset X$ and a family $\mathcal{B}_\alpha$ of open subsets of X such that the following conditions hold:

(a) $\mathcal{B}_\alpha$ is a pseudobase for F_α in X.

(b) $s(F_\alpha) \leq \mu$ and $|\mathcal{B}_\alpha| \leq \mu$.

(c) $F_\beta \subset F_\alpha$ for $\beta < \alpha$.

(d) If $\mathcal{K}$ is a finite subfamily in $\bigcup\{\mathcal{B}_\alpha: \alpha < \bar{\alpha}\}$, where $\bar{\alpha} < \alpha'$ and $X \setminus \bigcup \mathcal{K} \neq \emptyset$, then also $F_{\bar{\alpha}} \setminus \bigcup \mathcal{K} \neq \emptyset$.

Let us construct the corresponding set $F_{\alpha'}$ and family $\mathcal{B}_{\alpha'}$. Define $\mathcal{J}_{\alpha'} = \bigcup\{\mathcal{B}_\alpha: \alpha < \alpha'\}$ and for every set $U \in \bigvee \mathcal{J}_\alpha$, fix a point $x(U) \in X \setminus U$ (provided, of course, that $X \setminus U \neq \emptyset$). Further, let

$$S_{\alpha'} = \{x(U): U \in V\mathcal{J}_{\alpha'}, U \neq X\},$$

$$F_{\alpha'} = \operatorname{Cl}\left(\bigcup\{F_\alpha: \alpha < \alpha'\}\right) \cup \operatorname{Cl} S_{\alpha'}.$$

It is clear that

$$s(F_{\alpha'} = \sum\{s(F_\alpha): \alpha < \alpha'\} + |S_{\alpha'}|,$$

and, since

$$|S_{\alpha'}| \leq \left|\bigvee \mathcal{J}_{\alpha'}\right| = |\mathcal{J}_{\alpha'}| = \left|\bigcup\{\mathcal{B}_\alpha: \alpha < \alpha'\}\right| \leq \sum\{|\mathcal{B}_\alpha|: \alpha < \alpha'\},$$

it follows from (b) that

$$s(F_{\alpha'}) \leq \mu \cdot |\alpha'| + \mu \cdot |\alpha'| \leq \mu \cdot |\alpha'| \leq \mu \cdot \nu = \mu.$$

But then, by the initial assumption there exists a pseudobase $\mathcal{B}_{F_{\alpha'}}$ for $F_{\alpha'}$ in X such that $|\mathcal{B}_{F_{\alpha'}}| \leq \mu$. Let $\mathcal{B}_{\alpha'} = \mathcal{B}_{F_{\alpha'}}$. It is obvious from the construction that $F_{\alpha'}$ and $\mathcal{B}_{\alpha'}$ are the desired objects, and thus we can assume that for all $\alpha < \nu$ a set F_α and a family $\mathcal{B}_\alpha$ satisfying conditions (a)–(d) have been constructed.

We set $F = \bigcup\{F_\alpha: \alpha < \nu\}$ and show that $F = X$. Indeed, if there exists a point $\bar{x} \in X \setminus F$, then for every $\alpha < \nu$ there exists by (a) a set $B_\alpha \in \mathcal{B}_\alpha$ such that $F_\alpha \subset B_\alpha \subset X \setminus \{\bar{x}\}$. But by (c) the family $\{F_\alpha: \alpha < \nu\}$ is increasing and $t(X) < \operatorname{cf}(\nu) = \nu$, so, by Assertion 2, F is closed and hence compact in X. Consequently, the covering $\mathcal{P} = \{B_\alpha: \alpha < \nu\}$ of F contains a finite subcovering $\mathcal{P}'$. Clearly,

(8)As for every cardinal-valued function, we write $\psi(x, X)$ instead of $\psi(\{x\}, X)$.

$\bigcup \mathcal{P}' = U' \in \bigvee \mathcal{J}_{\bar\alpha}$ for some $\bar\alpha < \nu$, and since $\bar x \in X\backslash U' \neq \varnothing$, it follows from (d) (by the choice of $S_{\bar\alpha}$) that also $F\backslash U' \supset F_{\bar\alpha}\backslash U' \neq \varnothing$. A contradiction has been obtained, since $\mathcal{P}'$ is a covering of F and, hence, $\bigcup \mathcal{P}' = U' \supset F$. Thus $X = F$ and, consequently,

$$s(X) \le s\left(\bigcup\{F_\alpha : \alpha < \nu\}\right) \le \sum\{s(F_\alpha) : \alpha < \nu\} \le \mu \cdot \nu = \mu.$$

Similarly, $s(H) \le \mu$ for every closed set $H \subset X$, and we get that $\bar s(X) \le \mu$ from R2 or simply Assertion 1.

COROLLARY 2.1°. *If X is a compact T_1-space with countable tightness and $\psi(F, X) \le \aleph_1$ for every closed set $F \subset X$, then $\bar s(X) \le \aleph_1$.*

We single out the special case

COROLLARY 2.1. *The density of a perfectly normal compact Hausdorff space does not exceed $\aleph_1$.*

In the case of a totally ordered space X the condition $c(X) \le \aleph_0$ obviously implies the hereditary Lindelöf property for X and, consequently, the existence of a perfectly normal totally ordered compactification (see [71], [60]).

In this particular case a result analogous to Corollary 2.1 was obtained by M. E. Rudin [71], and also by Juhász and Hajnal [52] (concerning this see [63], [32], and [60]).

The density of every totally ordered perfectly normal compact Hausdorff space (or, which is equivalent, the density of every totally ordered space with the Suslin condition) does not exceed $\aleph_1$.

The following result is needed in order to use Theorem 2.1 in its full strength.

ASSERTION 2.1. *Let $\mathcal{A}$ be a family of subsets of X, and let $\psi(\mathcal{A}, X) = \sup\{\psi(A, X) : A \in \mathcal{A}\} = \tau$. If the set $\bigcup \mathcal{A}$ is closed in X and X is a compact T_1-space, then*

$$\psi\left(\bigcup \mathcal{A}, X\right) \le |\mathcal{A}| \cdot \psi(\mathcal{A}, X) \le |\mathcal{A}| \cdot \tau.$$

PROOF. For every $A \in \mathcal{A}$ we fix a pseudobase $\mathcal{B}_A$ for A in X such that $|\mathcal{B}_A| \le \tau$, let $\mathcal{B} = \bigcup\{\mathcal{B}_A : A \in \mathcal{A}\}$, and show that the family $\bigvee \mathcal{B}$ is a pseudobase for $F = \bigcup \mathcal{A}$ in X. Indeed, if $\bar x \in X\backslash F$, then for every set $A \in \mathcal{A}$ there is a set $B_A \in \mathcal{B}_A$ such that $A \subset B_A \subset X\backslash\{\bar x\}$. It is then clear that the covering $\mathcal{P} = \{B_A : A \in \mathcal{A}\}$ contains a finite subcovering $\mathcal{P}'$, because F is compact. Obviously, $\bigcup \mathcal{P}' \in \bigvee \mathcal{B}$ and, moreover, $F \subset \bigcup \mathcal{P}' \subset X\backslash\{\bar x\}$. But this means that $\bigvee \mathcal{B}$ is a pseudobase for F in X, and since $|\bigvee \mathcal{B}| = |\mathcal{B}| \cdot \aleph_0 \le |\mathcal{A}| \cdot \tau$, the proof is complete.

Taking the family $\mathcal{A}$ in the hypothesis of Assertion 2.1 to be the family $\{\{x\} : x \in F\}$ of singleton sets, we get

ASSERTION 2.1′. *If X is a compact T_1-space and F is a closed subset of X, then $\psi(F, X) \le |F| \cdot \psi(X)$.*

Obviously, $|Y| \le s(Y)^{\aleph_0}$ for every sequential subset of a T_2-space X (see, for example, [8]). Therefore, if X is a sequential compact Hausdorff space and F is closed in X, then, by Assertion 2.1′, $\chi(F,X) \leq s(F)^{\aleph_0} \cdot \chi(X)$ and, in particular, $\chi(F,X) \leq \chi(X)^{\aleph_0}$ when $s(F) \leq \chi(X)^{\aleph_0}$. Consequently, setting $\mu = \chi(X)^{\aleph_0}$ in the hypotheses of Theorem 2.1, we get a well-known theorem of Arkhangel′skiĭ [4] in the following form:

If X is a sequential compact Hausdorff space, then $s(X) \leq \chi(X)^{\aleph_0}$, and hence also $|X| \leq \chi(X)^{\aleph_0}$.

In particular, the cardinality of a compact Hausdorff space with the first axiom of countability does not exceed $\mathfrak{c}$ [4]. This theorem of Arkhangel′skiĭ was later re-proved by a whole series of authors (see [24], [18], [29], [58], [65], [70], [35], and [67]). Our proof of this result, as will be clear, consists of the proof of Theorem 2.1 along with insignificant additional arguments, and is thus very simple. Moreover, Theorem 2.1 provides at the same time the possibility of obtaining results new even in the class of perfectly normal compact Hausdorff spaces (Corollary 2.1). We remark that the approach of Pol [67], though very close to ours [35] (concerning this see [67]), nevertheless does not permit Corollary 2.1 to be proved.

In order to give a somewhat sharper estimate for cardinality (of compact spaces) with the help of Theorem 2.1, we need

ASSERTION 2.2. *Suppose that X is a compact T_1-space and $\psi(F,X) \leq \tau^\lambda$ for every closed $F \subset X$ such that $s(F) \leq \lambda$. If $t(X) < \lambda$, then $\psi(H,X) \leq \tau^\lambda$ for every closed $H \subset X$ such that $s(H) \leq \tau^\lambda$.*

PROOF. Suppose that $Y \subset X$, $\operatorname{Cl} Y = H$, and $|Y| < \tau^\lambda$. Setting $\mathcal{A} = \{\operatorname{Cl} A\colon A \subset Y,\ |A| \leq \lambda\}$, since $t(X) \leq \lambda$, we get that $H = \bigcup \mathcal{A}$. It is clear that $s([\operatorname{Cl} A]) \leq |A| \leq \lambda$ and, hence, $\psi(\operatorname{Cl} A,\ X) \leq \tau^\lambda$ for every $\operatorname{Cl} A \in \mathcal{A}$. Therefore, $\psi(H,X) \leq |\mathcal{A}| \cdot \tau^\lambda$, by Assertion 2.1, and, since $|\mathcal{A}| \leq |Y|^\lambda \leq \tau^\lambda$, we have that $\psi(H,X) \leq \tau^\lambda \cdot \tau^\lambda = \tau^\lambda$.

Because $\lambda^+ \leq \tau^\lambda$, Theorem 2.1 with $\mu = \tau^\lambda$ and Assertion 2.2 immediately give us the following result.

THEOREM 2.2. *Suppose that X is a compact T_1-space and $\psi(F,X) \leq \tau^\lambda$ for every closed set $F \subset X$ such that $s(F) \leq \lambda$. If $t(X) \leq \lambda$, then $\bar{s}(X) \leq \tau^\lambda$.*

COROLLARY 2.2. *If X is a compact T_1-space with countable tightness, and the pseudocharacter (in X) of each closed separable subset does not exceed $\mathfrak{c}$, then the density of X also does not exceed $\mathfrak{c}$.*

THEOREM 2.2′. *Let X be a compact T_1-space with the following property:*
(a) *If $F \subset X$ and $s(F) \leq t(X)$, then $|F| \leq \psi(X)^{t(X)}$.*
Then also $|X| \leq \psi(X)^{t(X)}$.

PROOF. By Assertion 2.1′, $\psi(F,X) \leq |F| \cdot \psi(X)$ for every closed set $F \subset X$, and, consequently, $\psi(F,X) \leq \psi(X)^{t(X)}$ by (a) if $s(F) \leq t(X)$. Therefore, setting $\tau = \psi(X)$ and $\lambda = t(X)$ in the hypotheses of Theorem 2.2, we get immediately that $s(X) \leq \tau^\lambda = \psi(X)^{t(X)}$. If we now take a set $Y \subset X$ such that $\operatorname{Cl} Y = X$ and $|Y| \leq \tau^\lambda$, we have that $X = \bigcup\{\operatorname{Cl} A\colon A \subset Y,\ |A| \leq \lambda\}$ and, consequently, by (a),

$$|X| \leq |Y|^\lambda \cdot \tau^\lambda \leq (\tau^\lambda)^\lambda \cdot \tau^\lambda = \tau^\lambda = \psi(X)^{t(X)}.$$

COROLLARY 2.2. *If X is a compact T_1-space with countable tightness, and the cardinality of every separable subset of X does not exceed $\mathfrak{c}$, then $|X| \leq \psi(X)^{\aleph_0}$.*

In the general case apparently the sharpest cardinality estimate analogous to that given above can be found in [35].

We now turn to the main problem of this section: that of clearing up the relation between the hereditary density $\bar{s}(X)$ and the hereditary cellularity $\bar{c}(X)$.

Almost without changing the proof, Theorem 2.1 can be strengthened somewhat, giving it a form needed in what follows.

THEOREM 2.1′. *Suppose that X is a compact T_1-space and that for every closed set $H \subset X$ such that $s(H) \leq \mu$ there exists a closed set $F_H \subset X$ such that $F_H \supset H$, $s(F_H) \leq \mu$, and $\psi(F_H, X) \leq \mu$. If $t(X) < \mu$, then $s(X) \leq \mu$.*

PROOF. It is clear that there exists a closed set $F \subset X$ such that $s(F) \leq \mu$ and $\psi(F, X) \leq \mu$. Let $F_0 = F$ and $\mathcal{B}_0 = \mathcal{B}_F$, where $\mathcal{B}_F$ is a pseudobase for F in X such that $|\mathcal{B}_F| \leq \mu$. Suppose that for all $\alpha < \alpha' < t(X)^+$ we have already constructed the corresponding F_α and $\mathcal{B}_\alpha$ satisfying conditions (a)–(d) in the proof of Theorem 2.1, and let the set $S_{\alpha'}$ be the same as there. Let

$$H_{\alpha'} = \operatorname{Cl}\left(\bigcup\{F_\alpha \colon \alpha < \alpha'\}\right) \cup \operatorname{Cl} S_{\alpha'}.$$

Since $s(H_{\alpha'}) \leq \mu$, by the initial conditions there exists a closed set $F_{H_{\alpha'}} \supset H_{\alpha'}$ and a pseudobase $\mathcal{B}$ for $F_{H_{\alpha'}}$ in X such that $s(F_{H_{\alpha'}}) \leq \mu$ and $|\mathcal{B}| \leq \mu$. Let $F_{\alpha'} = F_{H_{\alpha'}}$ and $\mathcal{B}_{\alpha'} = \mathcal{B}$. Repeating the proof of Theorem 2.1 word-for-word from this point, we get that $X = \bigcup\{F_\alpha \colon \alpha < \nu\}$ (where $\nu = t(X)^+$) and $s(X) \leq \mu$.

A number of subsequent assertions are needed to prove the main result in this section.

LEMMA 2.1. *Let $\mathcal{P}$ be an open covering of X. Then there exist a discrete set $D \subset X$ and a subfamily $\mathcal{P}' = \{P_x \colon x \in D\}$ such that $X = \operatorname{Cl} D \cup \bigcup \mathcal{P}'$ and $x \in P_x$ for all $x \in D$; moreover, it is clear that $|D| \leq \bar{c}(X)$ and $|\mathcal{P}'| \leq \bar{c}(X)$.*

PROOF. Let x_0 be an arbitrary point of X, and let P_0 be an element of $\mathcal{P}$ containing x_0. Suppose that for $\alpha < \alpha'$ we have already determined a point $x_\alpha \in X$ and a set P_α such that:

(a) $x_\alpha \in P_\alpha \in \mathcal{P}$, and

(b) $\{x_\alpha \colon \beta \leq \alpha < \alpha'\} \subset (X \setminus \bigcup\{P_\alpha \colon \alpha < \beta\}) \setminus \operatorname{Cl}\{x_\alpha \colon \alpha < \beta\}$ for all $\beta < \alpha'$.

If now there exists a point $\bar{x} \in G_{\alpha'} = (x \setminus \bigcup\{P_\alpha \colon \alpha < \alpha'\}) \setminus \operatorname{Cl}\{x_\alpha \colon \alpha < \alpha''\}$, then the corresponding point $x_{\alpha'}$ and set $P_{\alpha'}$ are obtained by setting $x_{\alpha'} = \bar{x}$ and $P_{\alpha'} = \overline{P}$ for some $\overline{P} \in \mathcal{P}$ such that $\bar{x} \in \overline{P}$. It is clear that for any α' the set $\{x_\alpha \colon \alpha \leq \alpha'\}$ thus constructed is discrete; indeed, letting

$$U_\beta = (X \setminus \operatorname{Cl}\{x_\alpha \colon \alpha < \beta\}) \cap P_\beta$$

we get from (a) and (b) that $U_\beta \cap \{x_\alpha \colon \alpha \le \alpha'\} = \{x_\beta\}$ for all $\beta \le \alpha'$. It is also obvious that for some $\bar\alpha$ (it is easy to see that $\bar\alpha < \bar c(X)^-$) the set $G_{\bar\alpha}$ is empty, and the discrete set $D = \{x_\alpha \colon \alpha < \bar\alpha\}$ and the family $\mathcal{P}' = \{P_\alpha \colon \alpha < \bar\alpha\}$ are the desired objects, since we then have $X = \operatorname{Cl} D \cup \bigcup\{P_\alpha \colon \alpha < \bar\alpha\}$.

LEMMA 2.2. *Suppose that X is a T_3-space and $A \subset X$. Then there exist sets G and H such that $A \subset G \subset H \subset X$ and*

$$\psi(G, X) \le \bar c(X), \qquad s(H) \le \bar c(X) s(A).$$

PROOF. By Lemma 2.1, for the covering $\mathcal{P} = \{P \colon P$ is open in X and $\operatorname{Cl} P \cap \operatorname{Cl} A = \varnothing\}$ of the set $U = X \backslash \operatorname{Cl} A$ there exist a subfamily $\mathcal{P}' \subset \mathcal{P}$ with $|\mathcal{P}'| \le \bar c(X)$ and a set $D \subset U$ such that $|D| \le \bar c(X)$ and $\operatorname{Cl} D \cup \bigcup \mathcal{P}' \supset X \backslash \operatorname{Cl} A$. Obviously, the sets $G = \bigcap\{X \backslash \operatorname{Cl} P \colon P \in \mathcal{P}'\}$ and $H = \operatorname{Cl} D \cup \operatorname{Cl} A$ are the desired ones.

ASSERTION 2.3. *Suppose that for each set $A \subset X$ such that $s(A) \le \mu$ there exist sets G and H such that $A \subset G \subset H \subset X$ and*

$$\psi(G, X) \le \mu, \qquad s(H) \le \mu.$$

If $t(X) < \mu$ and X is a compact Hausdorff space, then for each $A \subset X$ such that $s(A) \le \mu$ there exists a closed set $F_A \subset X$ such that $A \subset F_A$, $\psi(F_A, X) \le \mu$, and $s(F_A) \le \mu$.

PROOF. Fixing a corresponding G and H for a set A with $s(A) \le \mu$, we set $G_0 = G$ and $H_0 = H$. Suppose that for all $\alpha < \alpha' < \nu = t(X)^+$ we have already determined sets G_α and H_α such that the following conditions hold:

(a) $\psi(G_\alpha, X) \le \mu$, $s(H_\alpha) \le \mu$, and H_α is closed in X.

(b) $G_\alpha \subset H_\alpha$ and $H_\alpha \subset G_{\alpha+1}$ for $\alpha + 1 < \alpha'$.

Let $F_{\alpha'} = \operatorname{Cl}(\bigcup\{H_\alpha \colon \alpha < \alpha'\})$. Then, by (a),

$$s(F_{\alpha'}) \le \sum\{s(H_\alpha) \colon \alpha < \alpha'\} \le \mu \cdot \nu = \mu$$

and, consequently, there exist sets $G_{\alpha'}$ and $H_{\alpha'}$ such that $F_{\alpha'} \subset G_{\alpha'} \subset H_{\alpha'}$ and $\psi(G_{\alpha'}, X) \cdot s(H_{\alpha'}) \le \mu$; moreover, without loss of generality the set $H_{\alpha'}$ can be assumed to be closed. Constructing the corresponding sets H_α and G_α in this way for all $\alpha < \nu$, we get that $F_A = \bigcup\{H_\alpha \colon \alpha < \nu\}$.

$$F_A = \bigcup\{H_\alpha \colon \alpha < \nu\}.$$

By (b), the family $\mathcal{H} = \{H_\alpha \colon \alpha < \nu\}$ is increasing and $\nu = t(X)^+$, so, by Assertion 2, the set F_A is closed in X and, moreover, $s(F_A) \le \nu \cdot \mu = \mu$ by (a). If we now observe that the family $\mathcal{G} = \{G_\alpha \colon \alpha < \nu\}$ is also increasing, then we get that $\bigcup \mathcal{G} = \bigcup \mathcal{H} = F_A$ by (b), and, consequently, by Assertion 2.1 and (a),

$$\psi(F_A, X) \le |\mathcal{G}| \cdot \mu = \nu \cdot \mu = \mu.$$

Accordingly, F_A is the desired set.

We have now almost proved the main result in this section.

THEOREM 2.3°. *Suppose that X is a compact Hausdorff space. Then*

$$\bar{s}(X) \le t(X)^+ \cdot \bar{c}(X).$$

PROOF. Let $\mu = t(X)^+ \cdot \bar{c}(X)$. Then, by Lemma 2.2 and Assertion 2.3, for every $A \subset X$ with $s(A) \le \mu$ there exists a closed set $F_A \supset A$ in X such that $\psi(F_A, X) \cdot s(F_A) \le \mu$, and it remains only to apply Theorem 2.1′: considering that $t(X) < \mu$, we get that $s(X) \le \mu$. In precisely the same way, $s(H) \le t(H)^+ \cdot \bar{c}(H) \le \mu$ for every closed $H \subset X$ and, hence, $\bar{s}(X) \le \mu$ by R2 (or simply Assertion 1).

COROLLARY 2.3°. *If a compact Hausdorff space with countable tightness contains an uncountable discrete subset, then its hereditary density and cellularity are equal (i.e., $\bar{s}(X) = \bar{c}(X)$).*

In the case of a compact Hausdorff space we have $t(X) \le \bar{c}(X)$ by R1, so Theorem 2.3° immediately gives us

THEOREM 2.3. *Suppose that X is a compact Hausdorff space. Then $\bar{s}(X) \le \bar{c}(X)^+$.*

COROLLARY 2.3. *The hereditary density of a discrete-countable compact Hausdorff space does not exceed $\aleph_1$.*

The last theorem is a positive solution of a problem formulated in M. E. Rudin's book [72] (Problem 7, p. 65).

In the context of compact Hausdorff spaces we can write $\overline{\pi w}(X)$ in place of $\bar{s}(X)$ by R4. Hence we have

THEOREM 2.3′. *Suppose that X is a compact Hausdorff space. Then*

$$\overline{\pi w}(X) \le t(X)^+ \cdot \bar{c}(X) \le \bar{c}(X)^+.$$

As follows from results of Jech [56], Tennenbaum [78], and Jensen [57], the existence of a perfectly normal (and, hence, discrete-countable) nonseparable compact Hausdorff space is consistent (since a Suslin continuum is such a space) with the ZFC system of set theory axioms. Thus, Theorems 2.3°, 2.3, and 2.3′, like Corollary 2.1, cannot be strengthened with the tools of naive set theory by proving the equality $\bar{s}(X) = \bar{c}(X)$. (This equality constitutes a problem posed by de Groot in [49].)

At the same time, this equality is proved for the countable case with the help of additional axiomatic tools:

If X is a compact Hausdorff space, then $\bar{c}(X) \le \aleph_0$ implies $\bar{s}(X) \le \aleph_0$ [6] *under the premise $MA\&(\aleph_1 < \mathfrak{c})$.*

And although it is not clear whether one can prove the equality $\bar{c}(X) = \bar{s}(X)$ for all compact Hausdorff spaces under certain additional axiomatic premises, we nevertheless (since Corollary 2.3° asserts that in the class of compact Hausdorff spaces with countable tightness the inequality $\bar{c}(X) < \bar{s}(X)$ is endemic to the countable case: it is possible only when $\bar{c}(X) \le \aleph_0$) have the following improvement of Arkhangel′skiĭ's result above, from Corollary 2.3° and the result itself.

THEOREM 2.4. *Suppose that X is a compact Hausdorff space with countable tightness. Then $MA\&(\aleph_1 < \mathfrak{c})$ implies that $\bar{c}(X) = \bar{s}(X)$.*

This result is new even in the class of compact Hausdorff spaces with the first axiom of countability.

COROLLARY 2.4. *Under $MA\&(\aleph_1 < \mathfrak{c})$ the hereditary cellularity of a compact Hausdorff space with the first axiom of countability is equal to its hereditary density.*

Thus, the problem of de Groot has a positive solution in the class of compact Hausdorff spaces with countable tightness (and, in particular, with the first axiom of countability). Moreover, from R1, the obvious inequality $\bar{c}(X) \leq \bar{s}(X)$, and Theorem 2.3°, we get the sequence of inequalities

$$t(X) \leq \bar{c}(X) \leq \bar{s}(X) \leq t(X)^+ \cdot \bar{c}(X) \leq \bar{c}(X) \leq \bar{c}(X)^+,$$

which implies at once the following result if $\bar{c}(X) \neq \bar{s}(X)$.

THEOREM 2.5. *Suppose that X is a compact Hausdorff space. If $\bar{c}(X) \neq \bar{s}(X)$, then $\bar{c}(X) = t(X)$ and, moreover, $t(X)^+ = \bar{s}(X) = \bar{c}(X)^+$.*

COROLLARY 2.5. *The hereditary cellularity in a compact Hausdorff space is always equal to the hereditary density or to the tightness (i.e., $\bar{c}(X) = \bar{s}(X)$ or $\bar{c}(X) = t(X)$ always for compact Hausdorff spaces X).*

§3. Multiplicity of π-bases in compact Hausdorff spaces; d-density and tightness

The central result of the article is proved in this section.

THEOREM 3.1°. *Suppose that X is a compact Hausdorff space. Then for every $S \subset X$ there exists a family $\mathcal{I}(S)$ of open subsets of X which satisfies the following conditions*:

(a) *$\mathcal{I}(S)$ is a π-base for x in X for every point $x \in S$.*

(b) *The multiplicity of the family $\mathcal{I}(S)$ does not exceed $t(X)$.*

(c) *$|\mathcal{I}(S)| \leq |S| \cdot t(X)$.*

PROOF. We give a proof by cardinal number induction. For a cardinal τ let A_τ denote the following assertion:

A_τ: For every $S \subset X$ such that $|S| \leq \tau$ there exists a family $\mathcal{I}(S)$ satisfying conditions (a), (b), and (c).

By R3, for every $x \in X$ there exists a π-base $\mathcal{B}_x$ for x in X such that $|\mathcal{B}_x| \leq t(X)$. Therefore, for every set $S \subset X$ with $|S| \leq t(X)$ the family $\mathcal{I}(S) = \bigcup\{\mathcal{B}_x : x \in S\}$ satisfies conditions (a)–(c); hence assertion A_τ obviously holds for any $\tau \leq t(X)$ (in particular, for $\tau = \aleph_0$). Suppose now that $\nu > t(X)$ and that A_τ has been proved for all $\tau < \nu$. We show that A_ν then also holds.

Let us use transfinite induction to construct the corresponding family $\mathcal{I}(S)$ for a set $S \subset X$ of cardinality ν. We well-order S as follows:

$$S = \{x_\alpha : \alpha \text{ is not a limit ordinal and } \alpha < \nu\},$$

and assume that for all $\alpha < \alpha' \le \nu$ we have already constructed a compact set $F_\alpha \subset X$ and a family $\mathcal{G}_\alpha$ of open subsets of X such that the following conditions hold:

(0) $F_\alpha \subset F_\beta$, $\mathcal{G}_\alpha \subset \mathcal{G}_\beta$ for $\alpha < \beta < \alpha'$, and $x_\alpha \in F_\alpha$ if $\alpha < \alpha'$ is not a limit ordinal.

(1) $\mathcal{G}_\alpha$ is a π-base for x in X for all $x \in F_\alpha$, and $|\mathcal{G}_\alpha| \le |\alpha| \cdot t(X)$.

(2) The multiplicity of the family $\operatorname{Cl} \mathcal{G}_\alpha = \{\operatorname{Cl} G \colon G \in \mathcal{G}_\alpha\}$ does not exceed $t(X)$.

(3) If $G \in \mathcal{G}_\beta \backslash \mathcal{G}_\alpha$, then $F_\alpha \cap \operatorname{Cl} G = \varnothing$.

(4) if $\mathcal{K}$ is a finite family in $\operatorname{Cl} \mathcal{G}_\alpha$ and $\bigcap \mathcal{K} \neq \varnothing$, then also $F_{\alpha+1} \cap \bigcap \mathcal{K} \neq \varnothing$ (for $\alpha + 1 < \alpha'$). (Here F_0 is taken to be $\{x_0\}$, and $\mathcal{G}_0$ is taken to be a π-base for x_0 in X of cardinality $\le t(X)$.) We now construct the corresponding compact set $F_{\alpha'}$ and family $\mathcal{G}_{\alpha'}$. Assume first that α' is not a limit ordinal: $\alpha' = \delta + 1$ (and hence $\alpha' < \nu$). Then for every nonempty $H \in \bigwedge \operatorname{Cl} G_\delta$ such that $H \cap F_\delta = \varnothing$ we fix a point $x(H) \in H$ and let $S_\delta = \{x(H) \colon H \in \bigwedge \operatorname{Cl} \mathcal{G}_\delta$ and $H \cap F_\delta = \varnothing\}$. It is clear that $|S_\delta| \le |\bigwedge \mathcal{G}_\delta| = |\mathcal{G}_\delta| \le |\delta| \cdot t(X)$ (the last relation in view of (1)), and hence $\tau = |S_\delta| < \nu$, because $\alpha' < \nu$ and $t(X) < \nu$. But then assertion A_τ is valid by assumption. Consequently, there exists a family $\mathcal{I}(S_\delta)$ satisfying conditions (a)–(c), and $F_\delta \cap S_\delta = \varnothing$ by construction, so the family $\mathcal{I}_\delta = \{V_U \colon U \in \mathcal{I}(S_\delta)$, $\operatorname{Cl} U \cap F_\delta = \varnothing\}$ (where V_U is a nonempty open subset of X such that $\operatorname{Cl} V_U \subset U$) has the following properties:

(i_1) $\mathcal{I}_\delta$ is a π-base for x in X for every point $x \in S_\delta$, and $|\mathcal{I}_\delta| \le |S_\delta| \cdot t(X) \le |\delta| \cdot t(X) = |\alpha'| \cdot t(X)$;

(i_2) The multiplicity of the family $\operatorname{Cl} \mathcal{I}_\delta = \{\operatorname{Cl} U \colon U \in \mathcal{I}_\delta\}$ does not exceed $t(X)$.

(i_3) $F_\delta \cap \operatorname{Cl} U = \varnothing$ for every set $U \in \mathcal{I}_\delta$.

Next, we choose a π-base for $x_{\alpha'}$ in X such that $|\mathcal{B}_{x_{\alpha'}}| \le t(X)$ and $\operatorname{Cl}(\bigcup \mathcal{B}_{x_{\alpha'}}) \cap F_\delta = \varnothing$. (If $x_{\alpha'} \in F_\delta$, then we set $\mathcal{B}_{x_{\alpha'}} = \varnothing$.) Then, in view of ($\mathrm{i}_1$)–($\mathrm{i}_3$) and (1)–(3), the family $\mathcal{G}_{\alpha'} = \mathcal{G}_\delta \cup \mathcal{I}_\delta \cup \mathcal{B}_{x_{\alpha'}}$ and the set $F_{\alpha'} = F_\delta \cup \operatorname{Cl} S_\delta \cup \{x_{\alpha'}\}$ also satisfy conditions (1)–(3), and the choice of the set S_δ ensures condition (4). Obviously, condition (0) also holds. Accordingly, we have carried out the induction step in the case of an α' which is not a limit ordinal.

Suppose now that α' is a limit ordinal and $\alpha' \le \nu$. It is then clear that conditions (0), (1), (3), and (4) hold for the family $\mathcal{G}_{\alpha'} = \bigcup\{\mathcal{G}_\alpha \colon \alpha < \alpha'\}$ and the set $F_{\alpha'} = \operatorname{Cl}(\bigcup\{F_\alpha \colon \alpha < \alpha'\})$ by virtue of the corresponding induction hypotheses. For (1) it is necessary only to observe that

$$|\mathcal{G}_{\alpha'}| \le \sum\{|\mathcal{G}_\alpha| \colon \alpha < \alpha'\} \le t(X) \cdot \sum\{|\alpha| \colon \alpha < \alpha'\} \le |\alpha'|| \cdot t(X).$$

We show that (2) also holds. Indeed, this is obvious for $\operatorname{cf}(\alpha') \le t(X)$, because in this case $\operatorname{Cl} \mathcal{G}_{\alpha'}$ is a union of $\le t(X)$ families of multiplicity $\le t(X)$ in view of (0) and (2): $\operatorname{Cl} \mathcal{G}_{\alpha'} = \bigcup\{\operatorname{Cl} \mathcal{G}_\alpha \colon \alpha \in M\}$, where M is cofinal with α' and $|M| \le t(X)$. But if $\operatorname{cf}(\alpha') > t(X)$, then $F_{\alpha'} = \bigcup\{F_\alpha \colon \alpha < \alpha'\}$ by Assertion 2, since the family $\{F_\alpha \colon \alpha < \alpha'\}$ is nondecreasing by (0). Therefore, taking a family $\mathcal{H} \subset \operatorname{Cl} \mathcal{G}_{\alpha'}$ such that $\bigcap \mathcal{H} \neq \varnothing$, we have that $F_{\tilde{\alpha}} \cap \bigcap \mathcal{H} \neq \varnothing$ for some

$\bar{\alpha} < \alpha'$, since $F_{\alpha'} = \bigcup\{F_\alpha : \alpha < \alpha'\}$ is compact, and, by (4), $F_{\alpha'} \cap \bigcap \mathcal{K} \neq \varnothing$ for every finite subfamily $\mathcal{K} \subset \mathcal{H}$. But then $\mathcal{H} \subset \mathrm{Cl}\, \mathcal{G}_{\bar{\alpha}}$ by (3), and hence $|\mathcal{H}| \leq t(X)$ by (2) for every family $\mathcal{H} \subset \mathrm{Cl}\, \mathcal{G}_{\alpha'}$ such that $\bigcap \mathcal{H} \neq \varnothing$. But this means that the multiplicity of the family $\mathrm{Cl}\, \mathcal{G}_{\alpha'}$ is at most $t(X)$. Accordingly, we have completed the induction step also for a limit ordinal $\alpha' \leq \nu$, and for all $\alpha \leq \nu$ we can regard a family $\mathcal{G}_\alpha$ and a set F_α with conditions (0)–(4) as constructed. It remains to note that $S \subset F_\nu$ by condition (0), and the family $\mathcal{G}_\nu$ has properties (a), (b), and (c) according to conditions (1) and (2); consequently, it is the desired family: $\mathcal{G}_\nu = \mathcal{I}(S)$.

Assertion A_ν is thus true for any ν, and the proof is complete.

Setting $S = X$ in Theorem 3.1°, we immediately get the following result.

THEOREM 3.1. *Suppose that X is a compact Hausdorff space. Then X has a π-base whose multiplicity does not exceed $t(X)$.*

COROLLARY 3.1. *Every compact Hausdorff space with countable tightness has a point-countable π-base.*

This result is new in the class of sequential compact Hausdorff spaces and, moreover, in the class of compact Hausdorff spaces with the first axiom of countability; in the last case Theorem 3.1 admits a certain refinement. First, however, we need the simple

ASSERTION 3.1. *If $X = \mathrm{Cl}\{x \in X : \chi(x, X) \leq \tau\}$ and X has a π-base of multiplicity $\leq \tau$, then X also has a π-base $\mathcal{B}$ of multiplicity $\leq \tau$ that simultaneously is an external base for some dense $Y \subset X$.*

PROOF. Let $\mathcal{I}$ be a π-base for X such that (1) $|\mathcal{I}_x| \leq \tau$ for every point $x \in X$. By the condition for every $U \in \mathcal{I}$ there exists a point $y(U) \in U$ with a base $\mathcal{B}_{y(U)}$ such that (2) $|\mathcal{B}_{y(U)}| \leq \tau$, and (3) $\bigcup \mathcal{B}_{y(U)} \subset U$.

Clearly, the set $Y = \{y(U) : U \in \mathcal{I}\}$ is dense in X (since $\mathcal{I}$ is a π-base for X); hence, $\mathcal{B} = \bigcup\{\mathcal{B}_y : y \in Y\}$ is a π-base for X that simultaneously is an external base for Y. It remains to observe that $\mathcal{B}_x \subset \bigcup\{\mathcal{B}_{y(U)} : U \in \mathcal{I}_x\}$ by (3), and $|\mathcal{B}_x| \leq \tau$ by (2) and (1).

Accordingly, $\mathcal{B}$ is the desired π-base for X.

From Theorem 3.1 and Assertion 3.1 we now immediately get

THEOREM 3.1′. *Suppose that X is a compact Hausdorff space, $t(X) \leq \tau$, and $X = \mathrm{Cl}\{x \in X : \chi(x, X) \leq \tau\}$. Then X contains a dense subset having an external base in X whose multiplicity (in X) is at most τ.*

COROLLARY 3.1′. *If a compact Hausdorff space X of countable tightness contains a dense subset with the first axiom of countability, then X also contains a dense subset with a point-countable (in X external) base.*

In particular, every compact Hausdorff space with the first axiom of countability contains a dense subset with a point-countable (in X external) base.

REMARK 3.1. From the arguments used in proving Lemma 2 in Arkhangel′skiĭ's article [5] (see also [11]) it is easy to deduce the following fact.

F.3.1. *Suppose that X is a compact Hausdorff space such that every compact set $F \subset X$ with $s(F) \le t(X)$ satisfies the condition (ρ_X): if $H \subset F$ is compact, then*

$$H = \mathrm{Cl}\{x \in H \colon \chi(x, H) \le t(X)\}.$$

Then $F = \mathrm{Cl}\{x \in F \colon \chi(x, F) \le t(X)\}$ for every compact set $F \subset X$; in particular, $X = \mathrm{Cl}\{x \in X \colon \chi(x, X) \le t(X)\}$.

It follows easily from F.3.1 that every $\aleph_0$-exact [11] (in particular, every Corson([9])) and, under the premise $LH \equiv$ "$2^{\aleph_0} < 2^{\aleph_1}$", every sequential compact Hausdorff space has a dense set of points with the first axiom of countability [5], [11]. Therefore, Theorem 3.2 implies

COROLLARY 3.1″. *Every $\aleph_0$-exact compact Hausdorff space X (for example, a Corson compact Hausdorff space) and, under LH, every sequential compact Hausdorff space has a dense subset with a point-countable (in X external) base.*

In this connection recall that the existence of a point-countable base for "the whole of" a compact Hausdorff space implies that it is metrizable [20]. Moreover, it should be mentioned that even for Corson compact Hausdorff spaces it is not possible to strengthen Corollary 3.1″ by using naive means to prove the existence of a dense subset with a σ-point-finite (a condition which follows point-countability in the hierarchy of multiplicities) base (or even π-base). Indeed, the existence of a nonseparable Suslin continuum implies (by Corollary 1 in [25]) the existence of a nonseparable perfectly normal Corson compact Hausdorff space; on the other hand, in a hereditarily Lindelöf space every σ-point-finite π-base is countable [34], and this obviously implies separability.

In connection with these results and for completeness of the exposition we present the obvious

ASSERTION 3.2. *Let $\mathcal{B}$ be a family of open subsets of X, and $A = \{x \in X \colon |\mathcal{B}_x| \le \tau\}$. Then $\mathrm{Cl}_\tau A = A$, and, consequently, A is closed in X if $t(X) \le \tau$.*

ASSERTION 3.2′. *Suppose that X is a regular space, $\mathrm{Cl}\, Y = X$, and Y has a π-base of multiplicity $\le \tau$. If $\mathrm{Cl}_\tau Y = X$ (and, in particular, if $t(X) \le \tau$), then X also has a π-base of multiplicity $\le \tau$.*

It is clear that Theorem 1.1 in §1 is a corollary to Theorem 3.1. The next result makes it also possible to strengthen and generalize results in §2 on the basis of Theorem 3.1.

LEMMA 3.1. *Suppose that $\mathcal{B}$ is a family of nonempty open subsets of X and $|\mathcal{B}_x| < \tau$ for all $x \in X$. Then there exists a family $\mathcal{D}$ of discrete subsets of X such that $|\mathcal{D}| \le \tau$ and $\mathcal{B} = \mathcal{B}_{\bigcup \mathcal{D}}$([10]) (i.e., $B \cap (\cup \mathcal{D}) \neq \varnothing$ for every set $B \in \mathcal{B}$).*

PROOF. For every family $\mathcal{I}$ of nonempty open subsets of X we fix a discrete set $D = D(\mathcal{I})$ such that $\bigcup \mathcal{J} = \bigcup \mathcal{J}_{D(\mathcal{J})}$. Such a set exists. Indeed, suppose that

([9])A *Corson space* is defined to be a space that imbeds topologically in a Σ-product of closed intervals [79].

([10])Here and below, $\mathcal{B}_A = \{B \in \mathcal{B} \colon B \cap A \neq \varnothing\}$.

points x_α for all $\alpha < \alpha'$ have already been chosen, and

$$G_{\alpha'} = \bigcup \mathcal{J} \setminus \bigcup \{\bigcup \mathcal{J}_{x_\alpha} : \alpha < \alpha'\} \neq \varnothing.$$

Then take $x_{\alpha'}$ to be a point in $G_{\alpha'}$. Obviously, for some δ this procedure leads to a set $\{x_\alpha : \alpha < \delta\}$ for which $\bigcup \mathcal{J} = \bigcup\{\bigcup \mathcal{J}_{x_\alpha} : \alpha < \delta\}$; it is easily seen that the set $\{x_\alpha : \alpha < \delta\}$ is discrete, since $\{x_\alpha : \alpha < \delta\} \cap (\bigcup \mathcal{J}_{x_\beta}) = \{x_\beta\}$ for any $\beta < \delta$. Let $D_0 = D(\mathcal{B})$ and $\mathcal{B}_1 = \mathcal{B} \setminus \mathcal{B}_D$, and assume that the discrete sets D_α have already been determined for all $\alpha < \alpha' < \tau$. We set $\mathcal{B}_{\alpha'} = \mathcal{B} \setminus \bigcup\{\mathcal{B}_{D_\alpha} : \alpha < \alpha'\}$ and $D_{\alpha'} = D(\mathcal{B}_{\alpha'})$, and show that the family $\mathcal{D} = \{D_\alpha : \alpha < \tau\}$ thus constructed is the desired one. Indeed, if there exists a set

$$B' \in \mathcal{B} \setminus \mathcal{B}_{\bigcup \mathcal{D}} = \mathcal{B} \setminus \bigcup\{\mathcal{B}_{D_\alpha} : \alpha < \tau\} = \mathcal{B} \setminus \bigcup\{\tilde{\mathcal{B}}_\alpha : \alpha < \tau\},$$

where $\tilde{\mathcal{B}}_\alpha = \mathcal{B}_{D_\alpha}$, then $B' \in \mathcal{B}_\alpha$ for all $\alpha < \tau$. But $\bigcup \mathcal{B}_\alpha = \bigcup \tilde{\mathcal{B}}_\alpha$ by construction; hence for each point $x' \in B'$ there exists for any $\alpha < \tau$ a set $B_\alpha \in \tilde{\mathcal{B}}_\alpha$ such that $x' \in B_\alpha$. Noting now that $\tilde{\mathcal{B}}_\alpha \cap \tilde{\mathcal{B}}_\beta = \varnothing$ for any $\alpha \neq \beta$, we get at once that $|\mathcal{B}_{x'}| \geq \tau$. Accordingly, $\mathcal{B} = \bigcup\{\mathcal{B}_{D_\beta} : \beta < \tau\} = \mathcal{B}_{\bigcup \mathcal{D}}$, and the proof is complete.

For a family $\mathcal{B}$ of subsets of X we let $k(\mathcal{B}) = \sup\{|\mathcal{B}_x| : x \in X\}$, $\nabla k(\mathcal{B}) = \sup\{|\mathcal{B}_x|^+ : x \in X\} = \min\{\tau : \tau > |\mathcal{B}_x| \text{ for all } x \in X\}$, and $mk(B) = \min\{\tau : \mathcal{B} = \bigcup\{\mathcal{B}^\alpha : \alpha < \tau\}, \text{ and } \nabla k(\mathcal{B}^\alpha) \leq \tau \text{ for all } \alpha < \tau\}$. It is clear that $k(\mathcal{B}) \leq mk(\mathcal{B}) \leq \nabla k(\mathcal{B}) \leq k(\mathcal{B})^+$.

If now $\mathcal{B} = \mathcal{B}_A$, where $A \subset X$, then clearly $|\mathcal{B}| \leq |A| \cdot k(\mathcal{B})$. Therefore, taking A to be the set $\bigcup \mathcal{D}$ in Lemma 3.1 and noting that $|\mathcal{D}| \leq \bar{c}(X) \cdot mk(\mathcal{B}) \leq \bar{c}(X) \cdot \nabla k(\mathcal{B})$, we easily get the next result.

LEMMA 3.1′. *If $\mathcal{B}$ is a family of open subsets of X, then*

$$|\mathcal{B}| \leq \bar{c}(X) \cdot mk(\mathcal{B}) \leq \bar{c}(X) \cdot \nabla k(\mathcal{B}).$$

COROLLARY 3.2°. *Suppose that $\mathcal{B}$ is a family of open subsets of X.*

(1) *If $\mathcal{B}$ is point-countable, then $|\mathcal{B}| \leq \bar{c}(X) \cdot \aleph_1$, and if $\mathcal{B}$ is σ-point-finite, then $|\mathcal{B}| \leq \bar{c}(X)$.*

(2) *If X is a discrete-countable space (in particular, a hereditarily Lindelöf space), then $|\mathcal{B}| \leq mk(\mathcal{B}) \leq \nabla k(\mathcal{B})$.*

We remark that Corollary 3.2° is contained in [34] (Lemma 2.4) (and the proof of Lemma 3.1 in this paper coincides for $\bar{c}(X) \leq \aleph_0$ with the proof of Lemma 2.4 in [34]). For the case of σ-point-finite families Corollary 3.2° was obtained also by Velichko [13]. Moreover, for this case the result is not hard to derive from Arkhangel′skiĭ's paper [10] (see also [77]). However, in [13] (as also in [10] and [77]) the proof makes essential use of the point-finiteness of the families and does not go through even in the case of point-countable families.

We now single out a natural cardinal invariant that fits very harmoniously in our system of estimates and makes it possible to strengthen and generalize, in particular, the results in §2.

The *d-density* of a space X is defined to be the cardinal number $ds(X) = \min\{\tau : \text{there exists a family } \mathcal{D} \text{ of discrete subsets of } X \text{ such that } \operatorname{Cl} \bigcup \mathcal{D} = X \text{ and } |\mathcal{D}| \leq \tau\}$ (see [12] and [28]).

If $ds(X) \leq \aleph_0$, then we say that X is d-separable. It is clear that $ds(X) \leq s(X)$. Also trivial is

ASSERTION 3.3. *$s(X) \leq ds(X) \cdot \bar{c}(X)$ and $\bar{s}(X) \leq \overline{ds}(X) \cdot \bar{c}(X)$.*

THEOREM 3.2. *Let $\mathcal{B}$ be a π-base in X. Then $ds(X) \leq mk(\mathcal{B}) \leq \nabla k(\mathcal{B})$.*

PROOF. By assumption, $\mathcal{B} = \bigcup\{\mathcal{B}^\alpha \colon \alpha < \tau\}$ and $\nabla k(\mathcal{B}^\alpha) \leq \tau$ for all $\alpha < \tau = mk(\mathcal{B})$. Therefore, for each $\alpha < \tau$ there exists by Lemma 3.1 a family $\mathcal{D}_\alpha$ of discrete subsets of X such that $|\mathcal{D}_\alpha| \leq \tau$ and $\mathcal{B}^\alpha = \mathcal{B}^\alpha_{\bigcup \mathcal{D}_\alpha}$. Let $\mathcal{D} = \bigcup\{\mathcal{D}_\alpha \colon \alpha < \tau\}$. It is clear that $|\mathcal{D}| \leq \tau$ and $\mathrm{Cl} \bigcup \mathcal{D} = X$, because $\mathcal{B}$ is a π-base for X. But this means that $\mathcal{D}$ is the desired family of discrete subsets.

COROLLARY 3.2. *The d-density of a space with a point-countable base does not exceed $\aleph_1$. A space with a σ-point-finite base is d-separable.*

An immediate consequence of Theorems 3.1 and 3.2 is now

THEOREM 3.3. *Let X be a compact Hausdorff space. Then $\overline{ds}(X) \leq t(X)^+$ and, consequently, $\bar{s}(X) \leq t(X)^+ \cdot \bar{c}(X)$.*

COROLLARY 3.3. *The d-density of a compact Hausdorff space with countable tightness does not exceed $\aleph_1$. In particular, a compact Hausdorff space with the first axiom of countability contains a dense subset that is the union of $\aleph_1$ discrete spaces.*

Clearly, it is impossible to prove the d-separability of every (or even every Corson) compact Hausdorff space with the first axiom of countability without using additional axiomatic means, since, by the obvious Assertion 3.3 (and Theorem 1 in [25]), this would mean also the separability of every perfectly normal compact Hausdorff space.

Supplement

We announced Theorem 3.3 in [28]. The proof of Theorem 3.1 given in [25] is based on the possibility (established in [25]) of reducing compact Hausdorff spaces to special subspaces of Tychonoff cubes. Here we have given a direct proof of Theorem 3.1 which chronologically was the first proof. The same applies to Theorem 1.1. (A brief presentation of a direct proof of Theorem 1.1 was given in [21] (§3), and Corollary 1.1 was announced in [33].) §§1 and 2 of our article are close to Chapter 2 in [38], where, in particular, Theorem 2.3° is proved ($\bar{s}(X) \leq t(X)^+ \cdot \bar{c}(X)$) (see also [28]), although the inequality $\bar{s}(X) \leq \bar{c}(X)^+$ was proved by the same method in [35].

All the results listed above, which constitute the main contents of the paper, were reported in the seminar of Academician P. S. Aleksandrov between 1972 and 1976.

BIBLIOGRAPHY

1. Paul Alexandroff and Paul Urysohn [P. S. Aleksandrov and P. S. Uryson], *Mémoire sur les espaces topologiques compacts*, Verh. Kon. Akad. Wetensch. Amsterdam. Afd. Natuurk. (1) **14** (1929), no. 1, 1–96; rev. Russian transl. "Nauka", Moscow, 1971.

2. G. P. Amirdzhanov and B. È. Shapirovskiĭ, *On everywhere dense subsets of topological spaces*, Dokl. Akad. Nauk SSSR **214** (1974), 249–252; English transl. in Soviet Math. Dokl. **15** (1974).

3. A. V. Arkhangel′skiĭ, *An approximation of the theory of dyadic compact Hausdorff spaces*, Dokl. Akad. Nauk SSSR **184** (1969), 767–770; English transl. in Soviet Math. Dokl. **10** (1969).

4. ——, *On the cardinality of compact Hausdorff spaces with the first axiom of countability*, Dokl. Akad. Nauk SSSR **187** (1969), 967–970; English transl. in Soviet Math. Dokl. **10** (1969).

5. ——, *The Suslin number and cardinality. Characters of points in sequential compact Hausdorff spaces*, Dokl. Akad. Nauk SSSR **192** (1971), 255–258; English transl. in Soviet Math. Dokl. **12** (1971).

6. ——, *On compact Hausdorff spaces satisfying the Suslin condition hereditarily. Tightness and free sequences*, Dokl. Akad. Nauk SSSR **199** (1971), 1227–1230; English transl. in Soviet Math. Dokl. **12** (1971).

7. ——, *There is no "naive" example of a nonseparable sequential compact Hausdorff space with the Suslin property*, Dokl. Akad. Nauk SSSR **203** (1972), 983–985; English transl. in Soviet Math. Dokl. **13** (1972).

8. A. V. Arkhangel′skiĭ and V. I. Ponomarev, *Foundations of general topology in problems and exercises*, "Nauka", Moscow, 1974 (Russian); English transl., Hindustan, Delhi, and Reidel, Dordrecht, 1984.

9. ——, *On dyadic compact Hausdorff spaces*, Dokl. Akad. Nauk SSSR **182** (1968), 993–996; English transl. in Soviet Math. Dokl. **9** (1968).

10. A. V. Arkhangel′skiĭ, *The property of paracompactness in the class of perfectly normal locally compact spaces*, Dokl. Akad. Nauk SSSR **203** (1972), 1231–1234; English transl. in Soviet Math. Dokl. **13** (1972).

11. ——, *On some topological spaces that occur in functional analysis*, Uspekhi Mat. Nauk **31** (1976), no. 5(191), 17–32; English transl. in Russian Math. Surveys **31** (1976).

12. ——, *Compact Hausdorff spaces and unions of countable families of metrizable spaces*, Dokl. Akad. Nauk SSSR **232** (1977), 989–992; English transl. in Soviet Math. Dokl. **18** (1977).

13. N. V. Velichko, *On the cardinality of open coverings of topological spaces*, Fund. Math. **80** (1973), 271–282. (Russian)

14. B. A. Efimov, *On the cardinality of Hausdorff spaces*, Dokl. Akad. Nauk SSSR **164** (1965), 967–970; English transl. in Soviet Math. Dokl.**6** (1965).

15. ——, *Dyadic compact Hausdorff spaces*, Trudy Moskov. Mat. Obshch. **14** (1965), 211–247; English transl. in Trans. Moscow Math. Soc. **14** (1965).

16. ——, *Solution of some problems on dyadic compact Hausdorff spaces*, Dokl. Akad. Nauk SSSR **187** (1969), 21–24; English transl. in Soviet Math. Dokl. **10** (1969).

17. B. A. Efimov and G. I. Chertanov, *On two properties of the Suslin continuum*, Math. Balkanica **6** (1976), 47–54 (1978). (Russian)

18. A. A. Gryzlov, *On the cardinality of spaces*, Third Tiraspol Sympos. General Topology and its Applications, "Shtiintsa", Kishinev, 1973, p. 29. (Russian)

19. Thomas Jech, *Lectures in set theory, with particular emphasis on the method of forcing*, Lecture Notes in Math., Vol. 217, Springer-Verlag, 1971.

20. A. S. Mishchenko, *On spaces with a point-countable base*, Dokl. Akad. Nauk SSSR **144** (1962), 985–988; English transl. in Soviet Math. Dokl. **3** (1962).

21. V. I. Malykhin and B. È. Shapirovskiĭ, *Martin's axiom and properties of topological spaces*, Dokl. Akad. Nauk SSSR **213** (1973), 532–535; English transl. in Soviet Math. Dokl. **14** (1973).

22. V. I. Ponomarev, *On spaces co-absolute with metric spaces*, Uspekhi Mat. Nauk **21** (1966), no. 4(130), 101–132; English transl. in Russian Math. Surveys **21** (1966).

23. V. V. Fedorchuk, *On π-sets*, Dokl. Akad. Nauk SSSR **233** (1977), no. 6, 1060–1063; English transl. in Soviet Math. Dokl. **18** (1977).

24. V. I. Ponomarev, *On the cardinality of compact Hausdorff spaces with the first axiom of countability*, Dokl. Akad. Nauk SSSR **196** (1971), 296–298; English transl. in Soviet Math. Dokl. **12** (1971).

25. B. È. Shapirovskiĭ, *On tightness, π-weight, and related concepts*, Latvijas Valsts Univ. Zinātņ. Raksti=Latv. Gos. Univ. Uchen. Zap. **257** (1976), no. 2, 88–99. (Russian)

26. V. V. Fedorchuk, *On the cardinality of hereditarily separable compact Hausdorff spaces*, Dokl. Akad. Nauk SSSR **222** (1975), 302–305; English transl. in Soviet Math. Dokl. **16** (1975).

27. ____, *Fully closed mappings and the consistency of some theorems in general topology with the axioms of set theory*, Mat. Sb. **99(141)** (1976), 3–33; English transl. in Math. USSR-Sb. **28** (1976).

28. B. È. Shapirovskiĭ, Seventh All-Union Topology Conf., Abstracts of Reports, Inst. Mat. Akad. Nauk BSSR, Minsk, 1977, p. 205. (Russian)

29. D. V. Chudnovskiĭ, Sixth All-Union Topology Conf., Abstracts of Reports, "Metsniereba", Tbilisi, 1972, p. 129. (Russian)

30. N. A. Shanin, *Products of topological spaces*, Trudy Mat. Inst. Steklov. **24** (1948). (Russian).

31. B. È. Shapirovskiĭ, *On discrete subspaces of topological spaces. Weight, tightness, and Suslin number*, Dokl. Akad. Nauk SSSR **202** (1972), 779–782; English transl. in Soviet Math. Dokl. **13** (1972).

32. ____, *On the density of topological spaces*, Dokl. Akad. Nauk SSSR **206** (1972), 559–562; English transl. in Soviet Math. Dokl. **13** (1972).

33. ____, *Cardinal invariants and properties of topological spaces*, Sixth All-Union Topology Conf., Abstracts of Reports, "Metsniereba", Tbilisi, 1972, pp. 129–131. (Russian)

34. ____, *On separability and metrizability of spaces with the Suslin condition*, Dokl. Akad. Nauk SSSR **207** (1972), 800–803; English transl. in Soviet Math. Dokl. **13** (1972).

35. ____, *Canonical sets and character. Density and weight in compact Hausdorff spaces*, Dokl. Akad. Nauk SSSR **218** (1974), 58–61; English transl. in Soviet Math. Dokl. **15** (1974).

36. ____, *On spaces with the Suslin condition and the Shanin condition*, Mat. Zametki **15** (1974), 281–288; English transl. in Math. Notes **15** (1974).

37. ____, *On the π-character and the π-weight in compact Hausdorff spaces*, Dokl. Akad. Nauk SSSR **223** (1975), 799–802; English transl. in Soviet Math. Dokl. **16** (1975).

38. ____, *On cardinal topological invariants*, Author's Summary of Candidate's Dissertation, Moscow State Univ., Moscow, 1976. (Russian)

39. A. V. Archangelskij [Arkhangel′skiĭ], *On cardinal invariants*, General Topology and Its Relations to Modern Analysis and Algebra (Proc. Third Prague Sympos., 1971; J. Novák, Academia, Prague, and Academic Press, New York, 1972, pp. 37–46.

40. ____, *On hereditary properties*, General Topology and Appl. **3** (1973), 39–46.

41. David Booth, *Ultrafilters on a countable set*, Ann. Math. Logic **2** (1970/71), 1–24.

42. W. W. Comfort, *A survey of cardinal invariants*, General Topology and Appl. **1** (1971), 163–199.

43. R. Engelking, *Cartesian products and dyadic spaces*, Fund. Math. **57** (1965), 287–304.

44. ____, *Outline of general topology*, PWN, Warsaw, North-Holland, Amsterdam, and Wiley, New York, 1968.

45. R. Engelking and A. Pełczyński, *Remarks on dyadic spaces*, Colloq. Math. **11** (1963), 55–63.

46. S. P. Franklin, *Spaces in which sequences suffice*. II, Fund. Math. **61** (1967), 51–56.

47. Maurice Fréchet, *Les espaces abstracts*, Gauthier-Villars, Paris, 1928.

48. Andrew M. Gleason, *Projective topological spaces*, Illinois J. Math. **2** (1958), 482–489.

49. J. de Groot, *Discrete subspaces of Hausdorff spaces*, Bull. Acad. Polon. Sci. Sér. Sci. Math. Astr. Phys. **13** (1965), 537–544.

50. A. Hajnal and I. Juhász, *Discrete subspaces of topological spaces*, Nederl. Akad. Wetensch Proc. Ser. A **70**=Indag. Math. **29** (1967), 343–356.

51. ——, *Discrete subspaces of topological spaces*. II, Nederl. Akad. Wetensch. Proc. Ser. A **72**=Indag. Math. **31** (1969), 18–30.

52. ——, *Some remarks on a property of topological cardinal functions*, Acta Math. Acad. Sci. Hungar. **20** (1969), 25–37.

53. ——, *A consequence of Martin's axiom*, Nederl. Akad. Wetensch. Proc. Ser. A **74**= Indag. Math. **33** (1971), 457–463.

54. ——, *A consistency result concerning hereditarily α-Lindelöf spaces*, Acta Math. Acad. Sci. Hungar. **24** (1973), 307–312.

55. John R. Isbell, *Remarks on spaces of large cardinal number*, Czechoslovak Math. J. **14** (**89**) (1964), 383–385.

56. Tomáš [Thomas] Jech, *Non-provability of Souslin's hypothesis*, Comment. Math. Univ. Carolinae **8** (1967), 291–305.

57. R. B. Jensen, *Souslin's hypothesis is incompatible with $V = L$*, Notices Amer. Math. Soc. **15** (1968), 935 (abstract 68T-E31).

58. F. B. Jones, *The utility of empty inverse limits*, General Topology and Its Relations to Modern Analysis and Algebra (Proc. Third Prague Sympos., 1971; J. Novák, Academia, Prague, and Academic Press, New York, 1972, pp. 223–228.

59. I. Juhász, *Martin's axiom solves Ponomarev's problem*, Bull. Acad. Polon. Sci. Sér. Sci. Math. Astr. Phys. **18** (1970), 71–74.

60. ——, *Cardinal functions in topology*, Math. Centre Tracts, no. 34, Math. Centrum, Amsterdam, 1971.

61. ——, *On two problems of A. V. Archangel'skiĭ*, General Topology and Appl. **2** (1972), 151–156.

62. Georges [Đuro] Kurepa, *Ensembles ordonnés et ramifiés*, Thèse, Paris, 1935=Publ. Math. Univ. Beograd **4** (1935), 1–138.

63. ——, *On the cardinal number of ordered sets and of symmetrical structures in dependence on the cardinal numbers of its chains and antichains*, Glasnik Mat.-Fiz. Astr. Društvo Mat. Fiz. Hrvatske Ser. II **14** (1959), 183–203.

64. D. A. Martin and R. M. Solovay, *Internal Cohen extensions*, Ann. Math. Logic **2** (1970), 143–178.

65. S. Mrówka, *Some consequences of Archangelskii's theorem*, Bull. Acad. Polon. Sci. Ser. Sci. Math. Astr. Phys. **19** (1971), 373–376.

66. Jun-iti Nagata, *On a necessary and sufficient condition of metrizability*, J. Inst. Polytech. Osaka City Univ. Ser. A Math. **1** (1950), 93–100.

67. R. Pol, *Short proofs of two theorems on cardinality of topological spaces*, Bull. Acad. Polon. Sci. Ser. Sci. Math. Astr. Phys. **22** (1974), 1245–1249.

68. T. Przymusiński and F. D. Tall, *The undecidability of the existence of a non-separable normal Moore space satisfying the countable chain condition*, Fund. Math. **85** (1974), 291–297.

69. Judith Roitman, *A space of small spread without the usual properties*, Proc. Amer. Math. Soc. **43** (1974), 245–248.

70. Prabir Roy, *The cardinality of first countable spaces*, Bull. Amer. Math. Soc. **77** (1971), 1057–1059.

71. Mary Ellen Rudin, *Souslin's conjecture*, Amer. Math. Monthly **76** (1969), 1113–1119.

72. ——, *Lectures on set theoretic topology*, Conf. Board Math. Sci. Regional Conf. Ser. Math. no. 23, Amer. Math. Soc., Providence, R. I., 1975.

73. R. M. Solovay and S. Tennenbaum, *Iterated Cohen extensions and Souslin's problem*, Ann. of Math. (2) **94** (1971), 201–245.

74. M. Souslin [Suslin], Problème 3, Fund. Math. **1** (1920), 223.

75. R. M. Stephenson, Jr., *Discrete subsets of perfectly normal spaces*, Proc. Amer. Math. Soc. **34** (1972), 605–608.

76. Franklin D. Tall, *On the existence of non-metrizable hereditarily Lindelöf spaces with point-countable bases*, Duke Math. J. **41** (1974), 299–304.

77. ——, *The countable chain conditions versus separability—applications of Martin's axiom*, General Topology and Appl. **4** (1974), 315–339.

78. S. Tennenbaum, *Souslin's problem*, Proc. Nat. Acad. Sci. U.S.A. **59** (1968), 60–63.

79. E. Michael and M. E. Rudin, *A note on Eberlein compacts*, Pacific J. Math. **72** (1977), 487–495.

80. V. V. Fedorčuk [Fedorchuk], *A compact space having the cardinality of the continuum with no convergent sequences*, Math. Proc. Cambridge Philos. Soc. **81** (1977), 177–181.

81. G. P. Amirdzhanov, *Density, Suslin number, and tightness of topological spaces*, Dokl. Akad. Nauk SSSR **209** (1973), 265–268; English transl. in Soviet Math. Dokl. **14** (1973).

82. M. G. Tkachenko, *The behavior of cardinal invariants under the union operation for a chain of spaces*, Vestnik Moskov. Univ. Ser. I Mat. Mekh. **1978**, no. 4, 50–58; English transl. in Moscow Univ. Math. Bull. **33** (1978).

Translated by H. H. MCFADEN

Amer. Math. Soc. Transl.
(2) Vol. **134**, 1987

F. I. Karpelevich's Compactification Is Homeomorphic to a Ball

G. F. KUSHNER

§**1.** In this article we prove the following result for the Karpelevich compactification ([1], §13), which is in a certain sense a generalization of the Cayley-Klein interpretation of Lobachevsky space.

THEOREM. *The Karpelevich compactification* $\overline{\mathcal{E}}$ *of any noncompact symmetric Riemannian space* $\mathcal{E}$ *is homeomorphic to a ball* (*and the boundary* $\mathcal{T}$ *of* $\mathcal{E}$ *in it is homeomorphic to a sphere*).

The idea of the proof is as follows. Let G be a maximal connected group of motions of the space $\mathcal{E}$, U the stationary subgroup of the distinguished point x_0 of it, $\mathfrak{G}$ and $\mathfrak{U}$ their Lie algebras, $\mathfrak{L}$ the orthogonal complement of $\mathfrak{U}$ in $\mathfrak{G}$ (with respect to the inner product $\langle\cdot,\cdot\rangle$ generating the Riemannian metric of $\mathcal{E}$), $\mathfrak{H} \subset \mathfrak{L}$ some ordered Cartan subalgebra of $\mathcal{E}$, Π and

$$\mathfrak{W} = \{\mathfrak{h} \in \mathfrak{H} | \langle \alpha, \mathfrak{h}\rangle \geq 0 \text{ for any } \alpha \in \Pi\} \tag{1}$$

the corresponding system of simple roots and the corresponding Weyl chamber of this space, and B and $\overline{B}$ the open and closed unit balls in $\mathfrak{L}$, respectively ([1], §1). Let

$$V = B \cap \mathfrak{W} \quad \text{and} \quad W = \exp \mathfrak{W} \cdot x_0. \tag{2}$$

It is not hard to prove that V and W are *sweeping-out subspaces of the spaces* B *and* $\mathcal{E}$ (i.e., sections of the fiberings $B \to B/U$ and $\mathcal{E} \to \mathcal{E}U$ of B and $\mathcal{E}$ by the orbits of the group U), and their closures $\overline{V}$ and $\overline{W}$ are *sweeping-out subspaces of the ball* $\overline{B}$ *and the compactification* $\overline{\mathcal{E}}$.

1980 *Mathematics Subject Classification* (1985 *Revision*). Primary 53C21, 53C35, 54D35; Secondary 17B05, 17B20, 53C22.

Translation of Trudy Sem. Vektor. Tenzor. Anal. Vyp. **19** (1979), 95–111; MR **82b**: 53069.

To prove the theorem it suffices to construct a so-called *admissible homeomorphism between the sweeping-out subspaces of $\overline{B}$ and $\overline{\mathcal{E}}$*, i.e., a homeomorphism

$$\varsigma: \overline{V} \to \overline{W} \tag{3}$$

such that $\varsigma(V) = W$ and the stationary subgroups U_b and $U_{\varsigma(b)}$ of any point $b \in \overline{V}$ and of its image $\varsigma(b)$ coincide (for, as is easy to see, the formula

$$\overline{\varsigma}(\mathrm{Ad}(u)b) = u \cdot \varsigma(b) \qquad (u \in U,\ b \in \overline{V})$$

defines a homeomorphism $\overline{\varsigma}: \overline{B} \to \overline{\mathcal{E}}$ of $\overline{B}$ into $\overline{\mathcal{E}}$ in the case when ς is an admissible homeomorphism).

§**2.** Before passing to the construction of the homeomorphism $\varsigma: \overline{V} \to \overline{W}$ between the sweeping-out subspaces of $\overline{B}$ and $\overline{\mathcal{E}}$, we determine tests for it to be admissible.

Let us begin with the proof of a lemma and some heuristic considerations.

Let Π^* be an arbitrary (not necessarily proper) subsystem of the system Π of simple roots of $\mathcal{E}$ corresponding to the ordered Cartan subalgebra $\mathfrak{H}: \Pi^* \in 2^{\Pi}$, and let

$$\mathring{\mathfrak{W}}_{\Pi^*} = \{\mathfrak{h} \in \mathfrak{H} \mid \forall \alpha \in \Pi^*\ \langle \alpha, \mathfrak{h}\rangle = 0 \wedge \forall \alpha \in \Pi \setminus \Pi^*\ \langle \alpha, \mathfrak{h}\rangle > 0\}, \tag{4}$$
$$\mathfrak{W}_{\Pi^*} = \{\mathfrak{h} \in \mathfrak{H} \mid \forall \alpha \in \Pi^*\ \langle \alpha, \mathfrak{h}\rangle = 0 \wedge \forall \alpha \in \Pi \setminus \Pi^*\ \langle \alpha, \mathfrak{h}\rangle \geq 0\} \tag{5}$$

be the corresponding open and closed faces of the Weyl chamber $\mathfrak{W}$ (see (1)) of this space.

LEMMA 1. *For any $\Pi_1, \Pi_2 \in 2^{\Pi}$ and any $\mathfrak{h}_1 \in \mathring{\mathfrak{W}}_{\Pi_1}$, $\mathfrak{h}_2 \in \mathring{\mathfrak{W}}_{\Pi_2}$,*

$$U_{\mathfrak{h}_1} = U_{\mathfrak{h}_2} \Leftrightarrow \Pi_1 = \Pi_2.$$

PROOF. 1) Assume that there exists a root $\alpha \in \Pi_1 \setminus \Pi_2$, and take $u \in U$ such that the restriction $\mathrm{Ad}(u)|_{\mathfrak{H}}$ coincides with the symmetry s_α along the root α ([1], §1.13). Since $\langle \alpha, \mathfrak{h}_1\rangle = 0$ and $\langle \alpha, \mathfrak{h}_2\rangle > 0$, it follows that $u \in U_{\mathfrak{h}_1} \setminus U_{\mathfrak{h}_2}$; but this contradicts the assumption.

2) In view of the symmetry of the condition with respect to the vectors $\mathfrak{h}_1$ and $\mathfrak{h}_2$, it suffices to prove that $U_{\mathfrak{h}_1} \subseteq U_{\mathfrak{h}_2}$. Let $u \in U_{\mathfrak{h}_1}$ and

$$\Pi' = \{\alpha \in \Pi \mid \langle \alpha, \mathfrak{h}\rangle = 0\ \forall \mathfrak{h} \in \mathfrak{H} \cap \mathrm{Ad}(u)\mathfrak{H}\};$$

$\mathfrak{H}'$ denotes the linear span of Π', and $\tilde{\mathfrak{H}}$ denotes the orthogonal complement of $\mathfrak{H}'$ in $\mathfrak{H}$. By Lemma 11.2.1 in [1], there exists a $u^* \in U$ such that

$$\mathrm{Ad}(u^*)\mathfrak{H} = \mathrm{Ad}(u)\mathfrak{H} \quad \text{and} \quad \mathrm{Ad}(u^*)\tilde{\mathfrak{h}} = \tilde{\mathfrak{h}} \quad \text{for any } \tilde{\mathfrak{h}} \in \tilde{\mathfrak{H}}. \tag{6}$$

Since $\mathrm{Ad}(u^{-1}u^*)\mathfrak{H} = \mathfrak{H}$, $\mathfrak{h}_1, \mathfrak{h}_2 \in \mathring{\mathfrak{W}}_{\Pi_1}$, and $\mathrm{Ad}(u^{-1}u^*)\mathfrak{h}_1 = \mathfrak{h}_1$, it follows ([2], Chapter V, §3, no. 3, Proposition 1) that $\mathrm{Ad}(u^{-1}u^*)\mathfrak{h}_2 = \mathfrak{h}_2$, i.e., $\mathrm{Ad}(u)\mathfrak{h}_2 = \mathrm{Ad}(u^*)\mathfrak{h}_2$. It remains to prove that $\mathrm{Ad}(u^*)\mathfrak{h}_2 = \mathfrak{h}_2$; but this follows from (6), since it is not hard to prove that $\mathring{\mathfrak{W}}_{\Pi_1} \subset \tilde{\mathfrak{H}}$, and $\mathfrak{h}_2 \in \mathring{\mathfrak{W}}_{\Pi_1}$ by assumption. The lemma is proved.

Since

$$\mathring{\mathfrak{W}}_{\Pi^*} = \bigcap_{\alpha\in\Pi\setminus\Pi^*} (\mathfrak{W}_{\Pi^*} \setminus \mathfrak{W}_{\Pi^*\cup\{\alpha\}})$$

(see (4) and (5)), it is natural to assume that in the problem of stationary subgroups a role analogous to $\mathring{\mathfrak{W}}_{\Pi^*}$ is played for the sweeping-out subspaces $\overline{V}$ and $\overline{W}$ of $\overline{B}$ and $\overline{\mathcal{E}}$ by the sets(1)

$$\hat{V}_{\Pi^*} = \bigcap_{\alpha\in\Pi\setminus\Pi^*} (\overline{V}_{\Pi^*} \setminus \overline{V}_{\Pi^*\cup\{\alpha\}}), \quad \text{where } V_{\Pi^*} = V \cap \mathfrak{W}_{\Pi^*}, \tag{7}$$

$$\hat{W}_{\Pi^*} = \bigcap_{\alpha\in\Pi\setminus\Pi^*} (\overline{W}_{\Pi^*} \setminus \overline{W}_{\Pi^*\cup\{\alpha\}}), \quad \text{where } W_{\Pi^*} = \exp \mathfrak{W}_{\Pi^*} \cdot x_0. \tag{8}$$

And if, moreover, it is taken into account that for points

$$b \in V \cap \mathring{\mathfrak{W}}_{\Pi_1} \subset \hat{V}_{\Pi_1} \quad \text{and} \quad x \in \exp \mathring{\mathfrak{W}}_{\Pi_2} \cdot x_0 \subset \hat{W}_{\Pi_2},$$

the equality $U_b = U_x$ holds if and only if $\Pi_1 = \Pi_2$, then it is natural to expect that the following proposition holds.

PROPOSITION 1. *A homeomorphism $\varsigma: \overline{V} \to \overline{W}$ of sweeping-out subspaces of the closed ball $\overline{B}$ and the compactification $\overline{\mathcal{E}}$ is admissible if and only if for any subsystem Π^* of the system Π of simple roots of $\mathcal{E}$*

$$\varsigma(\hat{V}_{\Pi^*}) = \hat{W}_{\Pi^*}, \tag{9}$$

or, which is equivalent,

$$\varsigma(\overline{V}_{\Pi^*}) = \overline{W}_{\Pi^*}. \tag{10}$$

REMARK 1. The equivalence of conditions (9) and (10) follows from (7), (8) and the equalities

$$\overline{V}_{\Pi^*} = \bigcup_{\Pi^*\subseteq\Pi'\subseteq\Pi} \hat{V}_{\Pi'} \quad \text{and} \quad \overline{W}_{\Pi^*} = \bigcup_{\Pi^*\subseteq\Pi'\subseteq\Pi} \hat{W}_{\Pi'}.$$

A proof of Proposition 1 follows easily from Lemma 2.

LEMMA 2. *For any $\Pi_1, \Pi_2 \in 2^\Pi$ and any $\mathfrak{h} \in \hat{V}_{\Pi_1}$ and $x \in \hat{W}_{\Pi_2}$*

$$U_\mathfrak{h} = U_x \Leftrightarrow \Pi_1 = \Pi_2.$$

Lemma 2, in turn, follows from Lemmas 1 and 3.

LEMMA 3. *For any $\Pi^* \in 2^\Pi$ and $x \in \hat{W}_{\Pi^*}$ there exists a vector $\mathfrak{h}^* \in \mathring{\mathfrak{W}}_{\Pi^*}$ such that $U_{\mathfrak{h}^*} = U_x$.*

We prepare to prove Lemma 3.

Let T be *the set of unit vectors in the Weyl chamber* $\mathfrak{W}$ *of* $\mathcal{E}$, and let $\mathcal{E}'_\mathfrak{h}$ ($\mathfrak{h} \in T$) be *the component of the boundary of* $\mathcal{E}$ *in* $\overline{\mathcal{E}}$ *generated by* $\mathfrak{h}$ ([1], §§8, 12, and 13). Let

$$x'_0 = \lim_{t\to+\infty} \exp \mathfrak{h}t \cdot x_0 \quad \text{and} \quad \Pi' = \{\alpha \in \Pi | \langle\alpha, \mathfrak{h}\rangle = 0\}$$

(1)Note that the families $\{\hat{V}_{\Pi^*}\}$ and $\{\hat{W}_{\Pi^*}\}$ ($\Pi^* \in 2^\Pi$) are partitions of the spaces $\overline{V}$ and $\overline{W}$.

be the corresponding distinguished point and system of simple roots, and let $\overline{W}'_{\mathfrak{h}}$ be the corresponding sweeping-out subspace of the compactification $\overline{\mathcal{E}}'_{\mathfrak{h}}$ (see (2)). It is easy to prove that the following partition is valid:

$$\overline{W} = W \cup \left(\bigcup_{\mathfrak{h} \in T} \overline{W}'_{\mathfrak{h}} \right). \tag{11}$$

It is clear from (11) that an arbitrary point $x \in \overline{W}$ is determined by a tuple

$$(\mathfrak{h}, \mathfrak{h}', \dots, \mathfrak{h}^{(k-1)}; \mathfrak{h}^{(k)}) \qquad (0 \le k \le l-1), \tag{12}$$

where l is the rank of $\mathcal{E}$, $\mathfrak{h}, \dots, \mathfrak{h}^{(k-1)}$ are unit vectors, $\mathfrak{h} \in \mathfrak{W}$, and $\mathfrak{h}^{(i)} \in \mathfrak{W}^{(i)}$ $(1 \le i \le k)$, where $\mathfrak{W}^{(i)}$ is the Weyl chamber of the boundary component $\mathcal{E}^{(i)}$ of the space $\mathcal{E}^{(i-1)}$ corresponding to the vector $\mathfrak{h}^{(i-1)}$.

(Note that the point x can be reproduced from (12) as follows. Construct the points

$$x'_0 = \lim_{t \to +\infty} \exp \mathfrak{h} t \cdot x_0, \dots, x_0^{(k)} = \lim_{t \to +\infty} \exp \mathfrak{h}^{(k-1)} t \cdot x_0^{(k-1)}.$$

Then $x = \exp \mathfrak{h}^{(k)} \cdot x_0^{(k)}$.)

The next lemma can be proved by using the criterion for convergence of a sequence in the compactification $\overline{\mathcal{E}}$ ([1], §13).

LEMMA 4. *Suppose that* $x \in \overline{\mathcal{E}}'$, *where* $\mathcal{E}'$ *is the boundary component generated by a vector* $\mathfrak{h} \in T$, *with the system* Π' *of simple roots,* $\overline{W}'$ *is the sweeping-out subspace of the compactification* $\overline{\mathcal{E}}'$, *and* Π^* *is an arbitrary subsystem of the system* Π *of simple roots of* $\mathcal{E}$. *Then*

$$x \in \overline{W}_{\Pi^*} \Leftrightarrow \Pi^* \subseteq \Pi' \quad \textit{and} \quad x \in \overline{W}'_{\Pi^*},$$

or, what is equivalent,

$$x \in \hat{W}_{\Pi^*} \Leftrightarrow \Pi^* \subseteq \Pi' \quad \textit{and} \quad x \in \hat{W}'_{\Pi^*}.$$

The next result can be proved with the help of Lemma 4 by induction on the rank l of $\mathcal{E}$.

LEMMA 5. *Suppose that* $x \in \overline{W}$, $(\mathfrak{h}, \mathfrak{h}', \dots, \mathfrak{h}^{(k-1)}; \mathfrak{h}^{(k)})$ $(0 \le k \le l-1)$ *is a tuple of vectors determining it,*

$$\Pi^{(j)} = \{\alpha \in \Pi \mid \forall 0 \le i \le j-1 \, \langle \alpha, \mathfrak{h}^{(i)} \rangle = 0\} \qquad (0 \le j \le k+1) \tag{13}$$

$\Pi^* \in 2^{\Pi}$, *and* $u \in U$. *Then:*

a) $x \in \hat{W}_{\Pi^*} \Leftrightarrow \Pi^{(k+1)} = \Pi^*$, *and*

b) $u \cdot x = x \Leftrightarrow \forall 0 \le i \le k \, \mathrm{Ad}(u)\mathfrak{h}^{(i)} = \mathfrak{h}^{(i)}$.

PROOF OF LEMMA 3. 1) First of all we prove that the linear span of the vectors $\mathfrak{h}^{(i)}$ $(0 \le i \le k)$ contains a vector $\mathfrak{h}^*$ in $\mathring{\mathfrak{W}}_{\Pi^*}$. It is clear from (13) and Lemma 5a) that $\langle \alpha, h^* \rangle = 0$ for any $\alpha \in \Pi^*$ and any

$$\mathfrak{h}^* = \sum_{i=0}^{k} l_i \mathfrak{h}^{(i)}. \tag{14}$$

It remains to choose the coefficients l_i so that

$$\langle \alpha, \mathfrak{h}^* \rangle > 0 \quad \text{for any } \alpha \in \Pi \setminus \Pi^*. \tag{15}$$

We represent $\Pi \setminus \Pi^*$ in the form

$$\Pi \setminus \Pi^* = \bigcup_{0 \le j \le k} (\Pi^{(j)} \setminus \Pi^{(j+1)}).$$

It follows easily from the definition of (12) and from (13) that for any $\alpha \in \Pi^{(j)}$ $(0 \le j \le k)$

$$\langle \alpha, \mathfrak{h}^* \rangle = \sum_{i=j}^{k} l_i \langle \alpha, \mathfrak{h}^{(i)} \rangle,$$

i.e., the inner product does not depend on the coefficients $l_1, \dots, l_{j-1}$. It is clear that inequality (15) holds for any $\alpha \in \Pi^{(k)} \setminus \Pi^*$ if $l_k > 0$. Suppose now that the coefficients $l_k, \dots, l_{j+1}$ $(0 \le j \le k-1)$ are so chosen that (15) holds for any $\alpha \in \Pi^{(j+1)} \setminus \Pi^*$. Then it holds for any $\alpha \in \Pi^{(j)} \setminus \Pi^*$ if we take

$$l_j > \max_{\alpha \in \Pi^{(j)} \setminus \Pi^{(j+1)}} \left\{ - \sum_{i=j+1}^{k} l_i \langle \alpha, \mathfrak{h}^{(i)} \rangle / \langle \alpha, \mathfrak{h}^{(j)} \rangle \right\}.$$

2) It remains to prove that $U_{\mathfrak{h}^*} = U_x$. The inclusion $U_x \subseteq U_{\mathfrak{h}^*}$ follows from Lemma 5b) and (14). Proceeding to the proof of the inclusion $U_{\mathfrak{h}^*} \subseteq U_x$, we observe that, by Lemma 5a), $\langle \alpha, \mathfrak{h}^{(i)} \rangle = 0$ for any $\alpha \in \Pi^*$ and any $0 \le i \le k$.

We now take an arbitrary $0 \le i \le k$ and a positive number t_i such that

$$t_i \ge \max_{\alpha \in \Pi \setminus \Pi^*} \{ -\langle \alpha, \mathfrak{h}^{(i)} \rangle / \langle \alpha, \mathfrak{h}^* \rangle \}.$$

Obviously, $t_i \mathfrak{h}^* + \mathfrak{h}^{(i)} \in \overset{\circ}{\mathfrak{W}}_{\Pi^*}$. By applying Lemma 1 to it and the vector $\mathfrak{h}^*$ it is easy to see that $\mathrm{Ad}(u)\mathfrak{h}^{(i)} = \mathfrak{h}^{(i)}$ for any $u \in U_{\mathfrak{h}^*}$ and any $0 \le i \le k$, and thus $u \in U_x$ by Lemma 5b).

§**3.** A heuristic consideration of sweeping-out subspaces $\overline{W}$ of the compactifications $\overline{\mathcal{E}}$ of spaces $\mathcal{E}$ of small rank suggests that it is expedient to precede the construction of an admissible homeomorphism $\varsigma : \overline{V} \to \overline{W}$ by the construction of coverings of the spaces $\overline{V}$ and $\overline{W}$ chosen in such a way that the elements of the covering of $\overline{W}$ have a simpler structure than that of $\overline{W}$ itself (Figure 1). This can be done as follows. Let

$$\Pi = (\alpha_1, \dots, \alpha_l) \quad \text{and} \quad \Omega = (\omega_1, \dots, \omega_l) \tag{16}$$

be systems of simple roots and fundamental weights of $\mathcal{E}$, ordered in some way $(2\langle \omega_i, \alpha_j \rangle / \langle \alpha_j, \alpha_j \rangle = \delta_{ij})$, and let $s = (s(1), \dots, s(l))$ be an arbitrary permutation of the indices $1, \dots, l$: $s \in P_l$. Denote by $\mathfrak{W}_s$ the closed simplicial cone spanned by the vectors

$$e_{s,j} = \sum_{i=1}^{j} \omega_{s(i)} \qquad (1 \le j \le l), \tag{17}$$

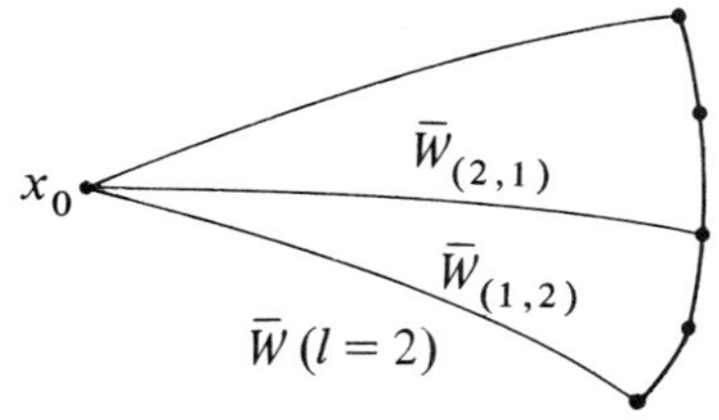

FIGURE 1

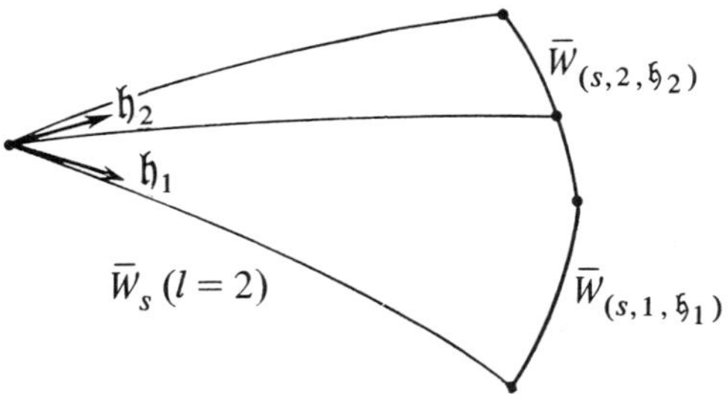

FIGURE 2

and let

$$V_s = V \cap \mathfrak{W}_s, \qquad W_s = \exp \mathfrak{W}_s \cdot x_0. \tag{18}$$

Obviously, the families $\{V_s\}$ and $\{W_s\}$ ($s \in P_l$) form coverings of the spaces V and W, and the families $\{\overline{V}_s\}$ and $\{\overline{W}_s\}$ ($s \in P_l$) form coverings of $\overline{V}$ and $\overline{W}$. We determine the structure of the subspace $\overline{W}_s$ ($s \in P_l$).

Let $T_{s,k}$ ($s \in P_l$, $1 \le k \le l$) be the set of unit vectors having an expansion

$$\mathfrak{h} = \sum_{j=1}^{k} h_j e_{s,j}, \quad \text{where } h_k > 0, \tag{19}$$

in the basis $e_{s,j}$ ($1 \le j \le l$) of the simplicial cone $\mathfrak{W}_s$, let $\mathfrak{H}'_{s,k}$ be the linear span of the roots $\alpha_{s(i)}$ ($k+1 \le i \le l$), and let $\omega'_{s(i)}$ and $e'_{s,i}$ ($k+1 \le i \le l$) be the projections on $\mathfrak{H}'_{s,k}$ of the vectors $\omega_{s(i)}$ and $e_{s,i}$.

It is known that to any vector $\mathfrak{h} \in T_{s,k}$ there corresponds a boundary component

$$\mathcal{E}'_{\mathfrak{h}} = \mathcal{E}(s, k, \mathfrak{h}) \tag{20}$$

with the distinguished point $x'_0 = \lim_{t\to+\infty} \exp \mathfrak{h}t \cdot x_0$.

It is easy to check that

$$\Pi'_{s,k} = (\alpha_{s(k+1)}, \dots, \alpha_{s(l)}) \quad \text{and} \quad \Omega'_{s,k} = (\omega'_{s(k+1)}, \dots, \omega'_{s(l)}) \tag{21}$$

are the ordered tuples of simple roots and fundamental weights for it corresponding to the permutation[2]

$$k(s) = (s(k+1), \dots, s(l)) \in P_{l-k}, \tag{22}$$

while the vectors $e'_{s,i}$ ($k+1 \le i \le l$), the subsequent closed simplicial cone $\mathfrak{W}'_{s,k}$ spanned by them, and its exponential

$$W'_{s,k,\mathfrak{h}} = \exp \mathfrak{W}'_{s,k} \cdot x'_0 \tag{23}$$

play the same role for this component and the permutation $k(s)$ as $e_{s,j}$ ($1 \le j \le l$), $\mathfrak{W}_s$, and W_s play for the space $\mathcal{E}$ and the permutation s.

---

(2)Note that $\Pi'_{s,k}$ and $\Omega'_{s,k}$ do not depend on the choice of the vector $\mathfrak{h} \in T_{s,k}$.

It is not hard to prove that $\overline{W}_s$ has the partition

$$\overline{W}_s = W_s \cup \left(\bigcup_{k=1}^{l} \left(\bigcup_{\mathfrak{h} \in T_{s,k}} \overline{W}'_{s,k,\mathfrak{h}} \right) \right). \tag{24}$$

Obviously, $T_{s,k}$ is homeomorphic to $[0,1)^{k-1}$; moreover, from (23) and the criterion for convergence of a sequence in the compactification $\overline{\mathcal{E}}$ ([1], §13) it follows easily that the spaces $\overline{W}_{s,k,\mathfrak{h}_1}$ and $\overline{W}_{s,k,\mathfrak{h}_2}$ are homeomorphic for any $\mathfrak{h}_1, \mathfrak{h}_2 \in T_{s,k}$; but then for any $1 \leq k \leq l$ the space $\bigcup_{\mathfrak{h} \in T_{s,k}} \overline{W}'_{s,k,\mathfrak{h}}$ is homeomorphic to $[0,1)^{k-1} \times \overline{W}_{s,k,\mathfrak{h}_0}$, where $\mathfrak{h}_0 \in T_{s,k}$ is a particular vector. Note also that

$$\dim \overline{W}_{s,k,\mathfrak{h}_0} = l - k,$$

and therefore

$$\dim \left(\bigcup_{\mathfrak{h} \in T_{s,k}} W'_{s,k,\mathfrak{h}} \right) = l - 1.$$

Thus, $\overline{W}_s$ actually has a simpler structure than $\overline{W}$; and an analysis of the structure of $\overline{W}_s$ for spaces of small rank leads one to suppose that it is homeomorphic to the closed l-dimensional cube (Figure 2).

§**4.** It is now possible at last to describe in general terms how to construct an admissible homeomorphism $\varsigma: \overline{V} \to \overline{W}$ of the sweeping-out subspaces of $\overline{B}$ and $\overline{\mathcal{E}}$.

First of all, homeomorphisms

$$\nu_s: \overline{V}_s \to \overline{Q}^l \quad \text{and} \quad \eta_s: Q^l \to \overline{W}_s \tag{25}$$

are constructed for an arbitrary permutation $s \in P_l$, where l is the rank of $\mathcal{E}$ and $\overline{Q}^l$ is the closed l-dimensional cube. Then their composition is taken:

$$\varsigma_s: \overline{V}_s \to \overline{W}_s. \tag{26}$$

The family $\{\varsigma_s\}$ $(s \in P_l)$ is compatible with the covering $\{\overline{V}_s\}$ $(s \in P_l)$ of $\overline{V}$ and thus gives rise to a continuous surjection

$$\varsigma: \overline{V} \to \overline{W}. \tag{27}$$

Next we observe that ς is a homeomorphism and is admissible.

§**5.** The homeomorphism $\nu_s: \overline{V}_s \to \overline{Q}^l$ is given by

$$\nu_s(b) = \begin{cases} \dfrac{|b|}{\mu(b)} (b^i)_{1 \leq i \leq l} \in \overline{Q}^l \setminus \{0\} & \text{if } b \in \overline{V}_s \setminus \{0\}, \\ 0 \quad \text{if } b = 0. \end{cases} \tag{28}$$

where $|b|$ is the modulus of b, b^i $(1 \leq i \leq l)$ are its coordinates in the basis $e_{s,i}$ of the simplicial cone $\mathfrak{W}_s$ (see (17)), and $\mu(b)$ is the maximal of these coordinates.

§6. We proceed to a description of the homeomorphism $\eta_s: \overline{Q}^l \to \overline{W}_s$. First of all, fix some *continuous function*

$$\delta(t) \qquad (0 \leq t < 1) \tag{29}$$

that is *monotonically increasing from* 0 *to* $+\infty$.

We introduce some notation and some auxiliary mappings. Let

$$Q^l = [0,1)^l, \qquad \overline{Q}^l = [0,1]^l. \tag{30}$$

We write

$$Q_{k,\dot{q}} = \{(\dot{q}, 1, q') \in \overline{Q}^l | q' \in \overline{Q}^{l-k}\}, \tag{31}$$

where $1 \leq k \leq l$ and $\dot{q} \in Q^{k-1}$. Obviously, $Q_{k,\dot{q}}$ is homeomorphic to the cube $\overline{Q}^{l-k}$, and the family

$$\{Q_{k,\dot{q}}\} \qquad (1 \leq k \leq l,\ \dot{q} \in Q^{k-1}) \tag{32}$$

is a partition of the set $\overline{Q}^l \setminus Q^l$.

For an arbitrary point $q \in Q^l$ let

$$\chi_s(q) = \sum_{i=1}^{l} \chi^i(q) e_{s,i}, \tag{33}$$

where

$$\chi^i(q) = \delta(q^i) \cdot \prod_{j=i+1}^{l} (1 + \delta(q^j)) \qquad (1 \leq i \leq l), \tag{34}$$

and $e_{s,i}$ $(1 \leq i \leq l)$ is a basis for the simplicial cone $\mathfrak{W}_s$. From (34) and (29) it follows easily that the mapping

$$\chi_s: Q^l \to \mathfrak{W}_s$$

is a continuous bijection. The same can be said of the mapping

$$Q^l \ni q \to \eta_s(q) = \exp \chi_s(q) \cdot x_0 \in W_s.$$

It remains to define the last mapping on the subspace $\overline{Q}^l \setminus Q^l$. With this aim we associate a vector $\tau_{s,k}(\dot{q}) \in T_{s,k}$ in a one-to-one and bicontinuous manner with each index $(k, \dot{q})$ of the partition (32) of this subspace (see §3), thereby associating with each $Q_{k,\dot{q}}$ the subspace

$$\overline{W}_{s,k,\dot{q}} = \overline{W}_{s,k,\tau_{s,k}(\dot{q})}, \tag{35}$$

and then use the induction hypothesis that the mapping η_s is known for all spaces $\mathcal{E}'$ with rank less than l.

The mapping

$$\tau_{s,k}: Q^{k-1} \to T_{s,k} \tag{36}$$

is constructed as follows. Take an arbitrary point $\dot{q} \in Q^{k-1}$ and let

$$\psi_{s,k}(\dot{q}) = \sum_{i=1}^{k-1} \psi_k^i(\dot{q}) e_{s,i} + e_{s,k}, \tag{37}$$

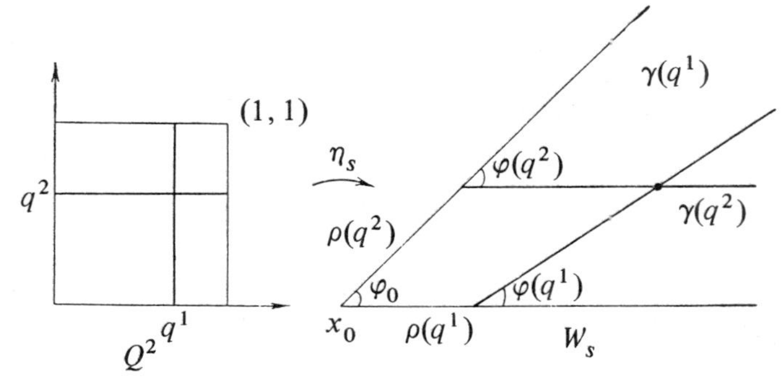

FIGURE 3

where

$$\psi_k^i(\dot q) = \delta(\dot q^i) \cdot \prod_{j=i+1}^{k-1} (1 + \delta(\dot q^j)) \qquad (1 \le i \le k-1) \tag{38}$$

and then normalize the vector $\psi_{s,k}(\dot q)$:

$$\tau_{s,k}(\dot q) = |\psi_{s,k}(\dot q)|^{-1} \cdot \psi_{s,k}(\dot q). \tag{39}$$

It is easy to prove that $\tau_{s,k}$ is a bijection and is, moreover, continuous.

Summing up, we get the formula

$$\eta_s(q) = \begin{cases} \exp \chi_s(q) \cdot x_0 \in W_s & \text{if } q \in Q^l, \\ \eta'_{s,k,\dot q}(q') \in \overline{W}'_{s,k,\dot q} & \text{if } q = (\dot q, 1, q') \in Q_{k,\dot q} \\ & (1 \le k \le l,\ \dot q \in Q^{k-1}) \end{cases} \tag{40}$$

for the mapping $\eta_s\colon \overline{Q}^l \to \overline{W}_s$. To determine the induction meaning of the mapping $\eta_s\colon \overline{Q}^l \to \overline{W}_s$ we consider the following

EXAMPLE. Let $l = 2$. Then (Figure 3) the vertical coordinate segment of the square Q^2 is mapped into a geodesic ray $\gamma(q^1)$ such that $\rho(q^1) \to +\infty$ and the angle $\varphi(q^1) \to 0$ as $q^1 \to 1$, while the horizontal coordinate segment is mapped into a geodesic ray $\gamma(q^2)$ such that $\rho(q^2) \to +\infty$ and the angle $\varphi(q^2)$ stays equal to the angle φ_0 as $q^2 \to 1$. The zero-dimensional subspaces $\overline{W}_{s,2,(q^1)}$ are exhausted by the singular points of the rays $\gamma(q^1)$; the singular points of the rays $\gamma(q^2)$ form the one-dimensional subspace $W_{s,1,\varnothing}$; and $\overline{W}_{s,1,\varnothing} = W_{s,1,\varnothing} \cup \eta_s(1,1)$ (see (24) and Figure 2).

Formula (40) arose as a natural generalization of this example.

PROPOSITION 2. *For any permutation $s \in P_l$ the mapping $\eta_s\colon \overline{Q}^l \to \overline{W}_s$ of the l-dimensional cube into the space $\overline{W}_s$ is a homeomorphism.*

For spaces $\mathcal{E}$ of rank 0 or 1 this proposition is valid, so its proof can be carried out by induction on the rank l of $\mathcal{E}$. Moreover, in view of the compactness of $\overline{Q}^l$ it suffices to prove that η_s is a continuous bijection.

§**7.** In order not to interrupt the proof of Proposition 2, we first formulate and prove two lemmas.

LEMMA 6. *Let* $\pi\colon \mathcal{E} \to \mathcal{E}'_*$ *be the projection of* $\mathcal{E}$ *onto its boundary component* $\mathcal{E}'_* = \mathcal{E}(s, k, \dot{q}_s)$ ([1], §8.6). *Then for any point* $(\dot{q}, q^k, q') \in Q^l$

$$\pi(\eta_s(\dot{q}, q^k, q')) = \eta_s(\dot{q}_*, 1, q') = \eta'_{s,k,\dot{q}_*}(q').$$

PROOF. Let $\mathfrak{H}' = \mathfrak{H}'_{s,k}$ be the Cartan subalgebra of the boundary component $\mathcal{E}'_*$ (see §3). By (33) and (34), and according to §3,

$$\mathrm{pr}_{\mathfrak{H}'}\, \chi_s(\dot{q}, q^k, q') = \sum_{i=k+1}^{l} \chi^i(q') e'_{s,i},$$

and thus, according to §8.6 in [1] and formula (40),

$$\pi(\eta_s(\dot{q}, q^k, q')) = \exp \sum_{i=k+1}^{l} \chi^i(q') e'_{s,i} \cdot x'_0 = \eta'_{s,k,\dot{q}_*}(q') = \eta_s(\dot{q}_*, 1, q').$$

LEMMA 7. *Let* $\pi\colon \mathcal{E}'_{\cdot} \to \overline{\mathcal{E}}'_*$ *be the projection of the boundary component* $\mathcal{E}'_{\cdot} = \mathcal{E}(s, m, \dot{q}_{\cdot})$ *into the closure of the boundary component*(3)

$$\mathcal{E}'_* = \mathcal{E}(s, k, \dot{q}_*) \qquad (1 \le k \le m \le l).$$

Then for any point $(\dot{q}, q^k, q') \in Q_{m,\dot{q}_0}$

$$\overline{\pi}(\eta_s(\dot{q}, q^k, q')) = \eta_s(\dot{q}_*, 1, q') = \eta'_{s,k,\dot{q}_*}(q').$$

PROOF. 1) Let $m = k$. To prove the lemma in this case it suffices to check the commutativity of the diagram

$$\begin{array}{ccc} & Q^{l-k} & \\ {}^{\eta'_{\cdot}}\swarrow & & \searrow^{\eta'_*} \\ W'_{\cdot} & \xrightarrow{\ \pi\ } & W'_*, \end{array} \tag{41}$$

where $W'_{\cdot}$ and W'_* are the sweeping-out subspaces of the boundary components $\mathcal{E}'_{\cdot}$ and $\mathcal{E}'_*$. Indeed, if this is so, then the diagram

$$\begin{array}{ccc} & \overline{Q}^{l-k} & \\ {}^{\eta'_{\cdot}}\swarrow & & \searrow^{\eta'_*} \\ \overline{W}'_{\cdot} & \xrightarrow{\ \overline{\pi}\ } & \overline{W}'_* \end{array}$$

is also commutative (since the mappings $\eta'_{\cdot}$ and η'_* in it are homeomorphisms by the induction hypothesis), and therefore

$$\overline{\pi}(\eta_s(\dot{q}, 1, q')) = \overline{\pi}(\eta'_{\cdot}(q')) = \eta'_*(q') = \eta_s(\dot{q}_*, 1, q')$$

by (40). As for the commutativity of diagram (41), it follows from (40), Lemma 6, and the commutativity of the diagram

$$\begin{array}{ccc} & \mathcal{E} & \\ {}^{\pi_{\cdot}}\swarrow & & \searrow^{\pi_*} \\ \mathcal{E}'_{\cdot} & \xrightarrow{\ \pi\ } & \mathcal{E}'_* \end{array}$$

(3)By footnote (2), $\mathcal{E}'_{\cdot} \infty \mathcal{E}'_*$ if $m = k$, and $\mathcal{E}'_{\cdot} \prec \mathcal{E}'_*$ if $m > k$. Hence we can speak of the projection $\pi\colon \mathcal{E}'_{\cdot} \to \overline{\mathcal{E}}'_*$ and its extension $\overline{\pi}\colon \overline{\mathcal{E}}'_{\cdot} \to \overline{\mathcal{E}}'_*$ ([1], §§10, 11, and 13.6).

(see [1], §10.3). Indeed, let $p^k \in [0,1)$. Then

$$\begin{aligned}\pi(\eta'_{\cdot}(q')) &= \pi(\eta_s(\dot{q},1,q')) = \pi\pi_{\cdot}(\eta_s(\dot{q},p^k,q')) \\ &= \pi_*(\eta_s(\dot{q},p^k,q')) = \eta_s(\dot{q}_*,1,q') = \eta'_*(q').\end{aligned}$$

2) Let $m > k$. By assumption, $(\dot{q}, q^k, q') \in Q_{m,\dot{q}_0}$; hence $q' = (\dot{q}', 1, q'')$, where $\dot{q}' \in Q^{m-k-1}$, $q'' \in \overline{Q}^{l-m}$, and $\dot{q}_{\cdot} = (\dot{q}, q^k, \dot{q}') \in Q^{m-1}$. Let $\mathfrak{H}'$ be the Cartan subalgebra of the boundary component $\mathcal{E}'_*$. According to (37), (38), and §3,

$$\operatorname{pr}_{\mathfrak{H}'} \psi_{s,m}(\dot{q}_0) = \sum_{i=k+1}^{m-1} \psi^i_m(\dot{q}')e'_{s,i} + e'_{s,m},$$

and thus the boundary component $\mathcal{E}''_*$ of $\mathcal{E}'_*$ onto which π maps $\mathcal{E}'_{\cdot}$ has the form

$$\mathcal{E}'_* = \mathcal{E}'_*(k(s), m-k, \dot{q}').$$

As in the preceding case, it suffices to see that the diagram

$$\begin{array}{ccccc} & & Q^{l-m} & & \\ & \swarrow^{\eta'_{\cdot}} & & \searrow^{\eta''_*} & \\ W'_{\cdot} & & \xrightarrow{\quad\pi\quad} & & W''_* \end{array} \tag{42}$$

is commutative, where $W'_{\cdot}$ and W''_* are sweeping-out subspaces of the spaces $E'_{\cdot}$ and E''_*.

By (40), Lemma 6, and the commutativity of the diagram

$$\begin{array}{ccc} \mathcal{E} & \xrightarrow{\pi_*} & \mathcal{E}'_* \\ \pi_{\cdot}\downarrow & & \downarrow\pi' \\ \mathcal{E}'_{\cdot} & \xrightarrow{\pi} & \mathcal{E}''_* \end{array}$$

(see [1], §11.4), the commutativity of diagram (42) can be proved as follows. Let $p^m \in [0,1)$. Then

$$\begin{aligned}\pi(\eta'_{\cdot}(q'')) &= \pi(\eta_s(\dot{q},q^k,\dot{q}',1,q'')) = \pi\pi_{\cdot}(\eta_s(\dot{q},q^k,\dot{q}',p^m,q'')) \\ &= \pi'\pi_*(\eta_s(\dot{q},q^k,\dot{q}',p^m,q'')) = \pi'(\eta_s(\dot{q}_*,1,\dot{q}',p^m,q'')) \\ &= \pi'(\eta'_*(\dot{q}',p^m,q'')) = \eta'_*(\dot{q}',1,q'') = \eta''_*(q'').\end{aligned}$$

§**8.** PROOF OF PROPOSITION 2. It is simple to prove that η_s is bijective. To prove continuity it suffices to observe that for any choice of the point $q_* \in \overline{Q}^l$ and a sequence $(q_n)_{n\in N}$ $(q_n \in \overline{Q}^l)$ convergent to it

$$\eta_s(q_n) \to \eta_s(q_*). \tag{43}$$

Without loss of generality we can assume that

$$q_* = (\dot{q}_*, 1, q'_*) \in Q_{k,\dot{q}_*} \qquad (1 \le k \le l,\ \dot{q}_* \in Q^{k-1}),$$

and consider two cases:

1°. $q_n = (\dot{q}_n, q^k_n, q'_n) \in Q^l$;

2°. $q_n = (\dot{q}_n, q^k_n, q'_n) \in Q_{m,\dot{q}_n}$ $(1 \le k \le m \le l)$

(see (30) and (31)).

Case 1°. Suppose that $\rho(x,y)$ is the Riemannian distance between points x and y in $\mathcal{E}$, γ_n is the geodesic ray with vertex x_0 passing through the point $\eta_s(q_n)$, $\gamma_* = \{\exp \tau_{s,k}(\dot{q}_*)t\cdot x_0 | t \geq 0\}$ is the geodesic ray generating the boundary component $\mathcal{E}'_* = \mathcal{E}(s,k,\dot{q}_*)$ whose closure contains the point $\eta_s(q_*)$, and $\pi\colon \mathcal{E} \to \mathcal{E}'_*$ is the projection.

To prove (43) it suffices to observe that

$$1)\ \rho(x_0, \eta_s(q_n)) \to +\infty; \quad 2)\ \gamma_n \to \gamma_*; \quad 3)\ \pi(\eta_s(q_n)) \to \eta_s(q_*)$$

(see [1], §13).

1) By (40), (33), and (34),

$$\rho(x_0, \eta_s(q_n)) = |\chi_s(q_n)| \geq |e_{s,k}| \cdot \chi^{(k)}(q_n) \geq |e_{s,k}| \cdot \delta(q_n^k),$$

but $q_n^k \to 1$, and thus $\delta(q_n^k) \to +\infty$ by (29).

2) It follows easily from (34) and the convergence $q_n^k \to 1$ that

$$\chi^i(q_n)/\chi^k(q_n) \to \begin{cases} \psi_k^i(\dot{q}_8) & \text{if } 1 \leq i \leq k-1, \\ 0 & \text{if } k+1 \leq i \leq l, \end{cases}$$

and thus $\chi_s(q_n)/\chi^k(q_n) \to \psi_{s,k}(\dot{q}_*)$. Hence, $\mathfrak{h}_n \to \mathfrak{h}_*$, where $\mathfrak{h}_n$ and $\mathfrak{h}_*$ are the unit vectors tangent to the geodesics γ_n and γ_* at x_0.

3) To prove the convergence $\pi(\eta_s(q_n)) \to \eta_s(q_*)$ it suffices to note that, by Lemma 6 and (40),

$$\pi(\eta_s(q_n)) = \eta'_{s,k,\dot{q}_*}(q'_n), \qquad \eta_s(q_*) = \eta'_{s,k,\dot{q}_*}(q'_*),$$

and to use the induction hypothesis.

Case 2° is handled similarly.

§**9.** To conclude the plan sketched in §4 it is necessary to see first of all that the family $\{\varsigma_s\}$ $(s \in P_l)$ of homeomorphisms is compatible with the covering $\{\overline{V}_s\}$ $(s \in P_l)$ of $\overline{V}$ and to observe that the continuous surjection $\varsigma\colon \overline{V} \to \overline{W}$ it gives rise to is injective. This is easily done by using the following lemmas.

LEMMA 8. *For any permutations $s, s_* \in P_l$ and any vectors $b \in \mathfrak{W}_s$ and $b_* \in \mathfrak{W}_{s_*}$, the equality $b = b_*$ holds if and only if in the expansions*

$$b = \sum_{i=1}^{l} b^i e_{s,i} \quad (b^i \geq 0) \quad \text{and} \quad b_* = \sum_{i=1}^{l} b_*^i e_{s_*,i} \quad (b_*^i \geq 0) \tag{44}$$

of these vectors in the bases of the simplicial cones $\mathfrak{W}_s$ and $\mathfrak{W}_{s_*}$ (see §3) $b^i = b_*^i$ for any $1 \leq i \leq l$ and $b^i = b_*^i = 0$ if $e_{s,i} \neq e_{s_*,i}$.

LEMMA 9. *For any choice of permutations $s, s_* \in P_l$ and any two vectors $b \in \overline{V}_s$ and $b_* \in \overline{V}_{s_*}$, the equality $b = b_*$ holds if and only if*

$$\nu_s(b) = \nu_{s_*}(b_*) \quad \text{and} \quad \nu_s(b)^i = \nu_{s_*}(b_*)^i = 0,$$

when $e_{s,i} \neq e_{s_*,i}$ (see §5).

LEMMA 10. *For any choice of permutations $s, s_* \in P_l$ and for any two points $q, q_* \in \overline{Q}^l$, the equality $\eta_s(q) = \eta_{s_*}(q_*)$ holds (see §6) if and only if*

$$q = q_* \quad \text{and} \quad q^i = q_*^i = 0 \quad \text{when } e_{s,i} \neq e_{s_*,i}. \tag{45}$$

PROOF OF LEMMA 8. Suppose that $e_{s,i} \neq e_{s_*,i}$ for some $1 \leq i \leq l$ (see (17)). We show first of all that there exist $1 \leq p \leq i$ and $1 \leq p_* \leq i$ such that

$$\omega_{s(p)} \notin \{\omega_{s_*(1)}, \ldots, \omega_{s_*(i)}\} \quad \text{and} \quad \omega_{s_*(p_*)} \notin \{\omega_{s(1)}, \ldots, \omega_{s(i)}\}. \tag{46}$$

Indeed, if this is not so, then

$$\{\omega_{s(1)}, \ldots, \omega_{s(i)}\} \subseteq \{\omega_{s_*(1)}, \ldots, \omega_{s_*(i)}\},$$

or conversely. But then these sets coincide, and thus $e_{s,i} = e_{s_*,i}$, which contradicts the assumption.

It follows from (46) that there exist $i \leq q_* \leq l$ and $i \leq q \leq l$ such that $\omega_{s(p)} = \omega_{s_*(q_*)}$ and $\omega_{s_*(p_*)} = \omega_{s(q)}$. Therefore, taking the inner products of both sides of the equality

$$\sum_{j=1}^{l} b^j e_{s,j} = \sum_{j=1}^{l} b_*^j e_{s_*,j} \tag{47}$$

first with $2\alpha_{s(p)}/\langle\alpha_{s(p)}, \alpha_{s(p)}\rangle$ and then with $2\alpha_{s(q)}/\langle\alpha_{s(q)}, \alpha_{s(q)}\rangle$, we get

$$\sum_{j=p}^{q-1} b^j + \sum_{j=q}^{l} b^j = \sum_{j=q_*}^{l} b_*^j; \qquad \sum_{j=q}^{l} b^j = \sum_{j=p_*}^{q_*-1} b_*^j + \sum_{j=q_*}^{l} b_*^j.$$

Hence,

$$\sum_{j=p}^{q-1} b^j = -\sum_{j=p_*}^{q_*-1} b_*^j, \tag{48}$$

and thus, by the nonnegativity, all the b^j and b_*^j in (48) are equal to zero; in particular, $b^i = b_*^i = 0$, because $p \leq i < q$ and $p_* \leq i < q_*$.

Accordingly, the only remaining terms in (47) are those for which $e_{s,j} = e_{s_*,j}$. The equalities $b^j = b_*^j$ not yet proved follow from the linear independence of the vectors $e_{s,j}$.

The converse assertion in Lemma 8 is obvious.

PROOF OF LEMMA 9. Since $b \in \overline{V}_s \in \mathfrak{W}_s$ and $b_* \in \overline{V}_{s_*} \subset \mathfrak{W}_{s_*}$, they can be represented in the form (44).

1) If $b = b_* = 0$, then the necessity of the statement in the lemma is obvious. Assume that $b = b_* \neq 0$. By (28), it suffices to prove that $b^i = b_*^i$ for any $1 \leq i \leq l$, and $b^i = b_*^i = 0$ if $e_{s,i} \neq e_{s_*,i}$; but this follows from Lemma 8.

2) Let $q = \nu_s(b) = \nu_{s_*}(b_*)$. If $q = 0$, then $b = b_* = 0$ by (28). Assume now that $q \neq 0$ and q^j is its maximal coordinate. It is clear from the same formula (28) and b^j and b_*^j are also maximal coordinates of the vectors b and b_*, i.e., $\mu(b) = b^j$ and $\mu_*(b_*) = b_*^j$, and thus $|b| = |b_*|$. Considering this along with the fact that $q^i = 0$ whenever $e_{s,i} \neq e_{s_*,i}$, we can assume that

$$\mu(b)^{-1} \cdot \sum_{i=1}^{l} b^i e_{s,i} = \mu_*(b_*)^{-1} \cdot \sum_{i=1}^{l} b_*^i e_{s_*,i}.$$

It remains to observe that $\mu(b) = \mu_*(b_*)$. Computing the squares of the lengths of the vectors in the last equality, we get that

$$\mu(b)^{-2}|b|^2 = \mu_*(b_*)^{-2}|b_*|^2.$$

By the equality $|b| = |b_*|$, what is necessary follows from this.

PROOF OF LEMMA 10. 1) Suppose that $\eta_s(q) = \eta_{s_*}(q_*) \in W$. Then $q, q_* \in Q^l$ and

$$\exp \chi_s(q) \cdot x_0 = \exp \chi_{s_*}(q_*) \cdot x_0 \in W_s \cap W_{s_*}.$$

Therefore $\chi_s(q) = \chi_{s_*}(q_*) \in \mathfrak{W}_s \cap \mathfrak{W}_{s_*}$. From this, according to (33) and Lemma 8, we get for any $1 \le i \le l$ that $\chi^i(q) = \chi^i(q_*)$ and $\chi^i(q) = \chi^i(q_*) = 0$ if $e_{s,i} \ne e_{s_*,i}$, and hence (45) holds by (34) and (29).

Suppose now that $\eta_s(q) = \eta_{s_*}(q_*) \in \overline{W} \setminus W$. Then, in view of (40),

$$q = (\dot q, 1, q') \in Q_{k,\dot q}, \qquad \eta_s(q) = \eta_{s,k,\dot q}(q') \in \overline{W}'_{s,k,\dot q},$$
$$q_* = (\dot q_*, 1, q'_*) \in Q_{k_*,\dot q_*}, \qquad \eta_{s_*}(q_*) = \eta_{s_*,k_*,\dot q_*}(q'_*) \in \overline{W}_{s_*,k_*,\dot q_*}.$$

The equality of the points $\eta_{s(q)}$ and $\eta_{s_*}(q_*)$ follows from the equality

$$\tau_{s,k}(\dot q) = \tau_{s_*,k_*}(\dot q_*),$$

from which, by (39), (37), and Lemma 8, we get

$$k = k_*, \qquad e_{s,k} = e_{s_*,k_*}; \tag{49}$$

$$\psi^i_k(\dot q) = \psi^i_k(\dot q_*), \quad \text{where } \psi^i_k(\dot q) = \psi^i_k(\dot q_*) = 0 \quad \text{if } e_{s,k} \ne e_{s_*,k_*}. \tag{50}$$

The validity of (45) for $1 \le i \le k-1$ follows from (50), (38), and (29).

To prove (45) for $k+1 \le i \le l$ it remains to note that $\{s(1), \dots, s(k)\} = \{s_*(1), \dots, s_*(k)\}$ by (49) and (17) so that

$$k(s) = (s(k+1), \dots, s(l)) \quad \text{and} \quad k_*(s_*) = (s_*(k+1), \dots, s_*(l))$$

are permutations of the same indices, and to use the induction hypothesis.

2) Let $q = q_* \in Q^l$, where $q^i = q^i_* = 0$ if $e_{s,i} \ne e_{s_*,i}$ $(1 \le i \le l)$. From this it follows easily (see (34)) that $\chi^i(q) = \chi^i(q_*)$, where $\chi^i(q) = \chi^i(q_*) = 0$ if $e_{s,i} \ne e_{s_*,i}$ $(1 \le i \le l)$. This leads to the equality of the vectors $\chi_s(q)$ and $\chi_{s_*}(q_*)$, and, consequently, also that of the points $\eta_s(q)$ and $\eta_{s_*}(q_*)$ (see (33) and (40)).

Assume now that $q = q_* \in \overline{Q}^l \setminus Q^l$, i.e.,

$$(\dot q, 1, q') = (\dot q_*, 1, q'_*) \in Q_{k,\dot q} \qquad (1 \le k \le l,\ \dot q \in Q^{k-1}).$$

The conditions (45) for the points $\dot q$ and $\dot q_*$ along with the equality $e_{s,k} = e_{s_*,k}$ imply equality of the vectors $\tau_{s,k}(\dot q)$ and $\tau_{s_*,k}(\dot q)$. The equality of the points $\eta_s(q) = \eta_{s,k,\dot q}(q')$ and $\eta_{s_*}(q_*) = \eta_{s_*,k,\dot q_*}(q'_*)$ now follows from the induction hypothesis.

§**10.** The admissibility of the homeomorphism $\varsigma\colon \overline{V} \to \overline{W}$ follows easily from the equality

$$\varsigma_s(V_{\Pi^*} \cap V_s) = W_{\Pi^*} \cap W_s, \tag{51}$$

which is valid for any $\Pi^* \subset 2^\Pi$ and $s \in P_l$, and from Proposition 1.

The proof of the theorem stated in §1 is complete.

REMARK 2. Proposition 2 gives a sufficiently clear representation of the topological structure of the subspaces $\overline{W}_s$ and $\overline{W}$ (see Figures 1 and 2), as does Proposition 1 of the fibering of the boundary $\mathcal{T}$ of the space $\mathcal{E}$ in the compactification $\overline{\mathcal{E}}$ by the orbits of the group U. The structure of $\mathcal{E}$ near the boundary $\mathcal{T}$ is also clear.

In conclusion the author takes it as a pleasant obligation to note that he was stimulated to work on this problem by talks with P. K. Rashevskiĭ and F. I. Karpelevich; he expresses his sincere appreciation to them.

BIBLIOGRAPHY

1. F. I. Karpelevich, *The geometry of geodesics and the eigenfunctions of the Beltrami-Laplace operator on symmetric spaces*, Trudy Moskov. Mat. Obshch. **14** (1965), 48–185; English transl. in Trans. Moscow Math. Soc. **14** (1965).

2. N. Bourbaki, *Groupes et algèbres de Lie*, Chaps. 4–6, Actualités Sci. Indust., no. 1337, Hermann, Paris, 1968.

Translated by H. H. MCFADEN

Amer. Math. Soc. Transl.
(2) Vol. **134**, 1987

A Remark on Sard's Theorem

V. A. BONDARENKO

In this note an assertion generalizing Sard's well-known theorem (see, e.g., [1] or [2]) is proved. It turns out that the conclusion of Sard's theorem remains valid in the case where the mapping is only differentiable (in the classical Sard theorem continuous differentiability is required).

Let G be an open set in $\mathbf{R}^n$, and let f be a mapping of G into $\mathbf{R}^n$ that is differentiable at each point. Recall that differentiability of f at a point $x \in G$ means the existence of a linear operator $f'(x)\colon \mathbf{R}^n \to \mathbf{R}^n$ such that

$$\|f(x+h) - f(x) - f'(x)h\| = O(\|h\|) \quad \text{for } x, x+h \in G.$$

Let $U = \{x \in G\colon \operatorname{rank} f'(x) < n\}$.

THEOREM. *The Lebesgue measure of the set $f(U)$ is zero.*

PROOF. The set G is the union of countably many n-dimensional cubes: $G = \bigcup_1^\infty Q_k$. Therefore it suffices to show that the measure of $f(U \cap Q)$ is zero for every cube Q contained in G. In turn, to prove this it suffices to establish that, for every $\varepsilon > 0$,

$$\overline{\operatorname{meas}} f(U \cap Q) \leq \varepsilon \operatorname{meas} Q \tag{1}$$

(the bar signifies Lebesgue outer measure).

Let $x \in U \cap Q$, and let $Q^l \subset Q$ $(l = 1, 2, \ldots)$ be a sequence of nested cubes with $x \in Q^l$ and diameters tending to zero. Then for some l_0

$$\overline{\operatorname{meas}} f(Q^{l_0}) \leq \varepsilon \operatorname{meas} Q^{l_0} \tag{2}$$

(for a proof see, e.g., [1]).

Now split Q into 2^n cubes of the same size, and denote them by Q_i^1 $(i = 1, 2, \ldots, 2^n)$. Next, split each Q_i^1 into 2^n cubes of the same size, and denote them by Q_i^2 $(i = 1, 2, \ldots, 2^{2n})$, etc.

1980 *Mathematics Subject Classification* (1985 *Revision*). Primary 57R70.
Translation of Vestnik Yaroslav. Univ. Vyp. 12 (1975), 11–12; MR **56** #9553.

Denote by Q^1_ε the union of all cubes Q^1_i for which

$$\overline{\text{meas}} f(Q^1_i) \leq \varepsilon \,\text{meas}\, Q^1_i.$$

Next, denote by Q^2_ε the union of all cubes Q^2_i disjoint from Q^1_ε for which

$$\overline{\text{meas}} f(Q^2_i) \leq \varepsilon \,\text{meas}\, Q^2_i.$$

Proceeding analogously, construct sets Q^3_ε, etc.

Consider the set $U_\varepsilon = \bigcup_1^\infty Q^k_\varepsilon$. From (2) it follows that $U \cap Q \subset U_\varepsilon$. On the other hand it is obvious that

$$\overline{\text{meas}} f(U_\varepsilon) \leq \varepsilon \,\text{meas}\, U_\varepsilon \leq \varepsilon \,\text{meas}\, Q.$$

From this follows the required inequality (1).

The theorem is proved.

The theorem remains valid if the mapping f is not differentiable on all of G. The proof given above also covers this case.

BIBLIOGRAPHY

1. Georges de Rham, *Variétés différentiables. Formes, courants, formes harmoniques*, Actualités Sci. Indust., no. 1222, Hermann, Paris, 1955.

2. J. T. Schwartz, *Differential geometry and topology*, Gordon and Breach, New York, 1968.

Translated by B. SILVER

Amer. Math. Soc. Transl.
(2) Vol. **134**, 1987

On the Equivalence of Exact Morse Functions. I

UDC 513.836

V. V. SHARKO

In this paper we investigate exact Morse functions on smooth simply connected manifolds of dimension greater than 5. By a *regular Morse function* on a manifold M^n is meant a function $f: M^n \to \mathbf{R}^1$ which has only nondegenerate critical points, and such that if m is a critical point then $f(m) =$ the index of m (see [1]). It is well known that in this case $M_k^f = f^{-1}[0, k + \frac{1}{2}]$ is a submanifold of M^n $(0 \leq k \leq n)$ and $M_k^f = M_{k-1}^f \cup_{\psi_1} (D^k \times D^{n-k}) \cdots \cup_{\psi_p} (D^k \times D^{n-k})$, where $\psi_i: \partial D^k \times D^{n-k} \to f^{-1}[k - \frac{1}{2}]$ are embeddings.

A regular Morse function is called *exact* if the number of critical points of index λ is equal to

$$N_\lambda = p_\lambda + q_\lambda + q_{\lambda-1},$$

where p_λ is the rank of $H_\lambda(M^n, \mathbf{Q})$ and q_λ is the minimal number of generators of $\operatorname{Tors}(H_\lambda(M^n, \mathbf{Z}))$ (see [1]).

DEFINITION 1. Exact Morse functions f and g on M^n are called *equivalent* if there exists a diffeomorphism $h: M^n \to M^n$ isotopic to the identity and such that $h(M_k^f) = M_k^g$ for $0 \leq k \leq n$.

Agoston proved that if the groups $H_i(M^n, \mathbf{Z})$ have no torsion, then all exact Morse functions are equivalent. In the case when the groups $H_i(M^n, \mathbf{Z})$ have torsion, he constructed an example of a manifold and two inequivalent exact Morse functions on it. He also proved that a diffeomorphism acting as the identity on the torsion of the $H_i(M^n, \mathbf{Z})$ carries an exact Morse function into one equivalent to it (see [2]).

In this paper we give necessary and sufficient conditions for the equivalence of exact Morse functions. To each exact Morse function there is associated, in

1980 *Mathematics Subject Classification* (1985 *Revision*). Primary 57R65, 57R70.

Translation of Studies in Function Theory and Topology, Preprint No. 12, Inst. Mat. Akad. Nauk Ukrain. SSR, Kiev, 1979, pp. 14–26; MR **81k**:57032.

a canonical way, a residue in $\mathrm{Tors}(H_\lambda(M^n, \mathbf{Z}))$. In order that the functions be equivalent, it is necessary and sufficient that these residues coincide.

We now make an algebraic digression. Let G be a finitely generated abelian group, and $\mu(G) = k+l$ the minimal number of its generators. We are interested in epimorphisms of free abelian groups of rank $k+l$ onto G.

DEFINITION 2. Let $\varphi_i\colon F_i \to G$ be epimorphisms, where F_i is a free abelian group of rank $k+l$ $(i = 1, 2)$. We shall say that φ_i *is equivalent to* φ_2 if there exists an isomorphism $L\colon F_1 \to F_2$ such that $\varphi_1 = \varphi_2 \circ L$.

Let $g_1, \dots, g_k, \overline{g}_1, \dots, \overline{g}_l$ be a minimal system of generators of the group G, where the g_i are free generators $(1 \le i \le k)$ and the $\overline{g}_j$ are generators of the torsion such that the order of $\overline{g}_j$ divides the order of $\overline{g}_{j+1}$. We denote by m_j the order of $\overline{g}_j$.

Suppose given an epimorphism $\varphi\colon F \to G$, where F is a free group of rank $k+l$. Choose a basis $(f_1, \dots, f_k, f_{k+1}, \dots, f_l)$ in F. Then it is obvious that $\varphi(f_s)$ can be uniquely represented in the form

$$\varphi(f_s) = \sum_{i=1}^{k} a_{is} g_i + \sum_{j=1}^{l} a_{k+j,s}\overline{g}_j,$$

where $a_{is} \in \mathbf{Z}$ and $a_{k+j,s}$ is a residue modulo m_j $(1 \le s \le k+l)$. We form

$$A(\varphi) = \begin{pmatrix} a_{11} & \dots & a_{1k} & a_{1,k+1} & \dots & a_{1,k+l} \\ \vdots & & & & & \vdots \\ a_{k+l,1} & & \dots & \dots & & a_{k+l,k+l} \end{pmatrix}$$

in which in the first k columns the entries are integers, and in the $(k+j)$th column $(1 \le j \le l)$ the entries are residues modulo m_j. It is clear that elementary transformations of bases in F correspond to elementary row operations of matrices. Notice that in the first k columns the action is carried out over $\mathbf{Z}$, and for the remaining columns with the corresponding residues.

We have the following lemma.

LEMMA 1. *Let $\varphi\colon F \to G$ be an epimorphism, and F a free abelian group of rank $k+l$. Then in F there exists a basis $f_1, \dots, f_k, f_{k+1}, \dots, f_{k+l}$ such that φ is represented in the form*

$$\varphi(f_i) = g_i \quad (1 \le i \le k), \qquad \varphi(f_{k+1}) = \alpha\overline{g}_1, \qquad \varphi(f_j) = \overline{g}_j \quad (2 \le j \le l),$$

where α is a residue modulo m_1, $(\alpha, m_1) = 1$, and $\alpha < m_1/2$.

We call α the *residue* of the epimorphism φ, and have the following theorem.

THEOREM 1. *Let $\varphi_t\colon F_t \to G$ $(t = 1, 2)$ be epimorphisms of free abelian groups of rank $k+l$ onto G. Then φ_1 is equivalent to φ_2 if and only if their residues coincide.*

Thus with each epimorphism we associate a residue. This fact is used in the sequel.

PROOF. *Sufficiency.* If φ_1 and φ_2 have the same residues, then obviously they are equivalent.

Necessity. We give a proof by contradiction. On the basis of the lemma, we choose in F_t a basis $f_1^t, \dots, f_k^t, f_{k+1}^t, \dots, f_{k+l}^t$ so that φ_t can be represented in the form

$$\begin{aligned} \varphi_t(f_i^t) &= g_i \qquad (1 \le i \le k), \\ \varphi_t(f_{k+1}^t) &= \alpha^t \overline{g}_{k+1}, \\ \varphi_t(f_{k+j}^t) &= \overline{g}_{k+j} \qquad (2 \le j \le l), \end{aligned}$$

and suppose $\alpha^1 \neq \alpha^2$.

Since φ_i is equivalent to φ_2, there exists an isomorphism $L\colon F_1 \to F_2$ so that $\varphi_1 = \varphi_2 \circ L$. Obviously, for the chosen bases L has the form

$$\begin{aligned} L(f_i^1) &= f_i^2 + \sum_{j=1}^{l} a_{ij} f_j^2 \qquad (1 \le i \le k), \\ L(f_{k+1}^1) &= (\beta + a_{k+1,k+1}) f_{k+1}^2 + \sum_{j=2}^{l} a_{k+1,k+j} f_j^2, \\ L(f_{k+j}^1) &= (1 + a_{k+j,k+j}) f_{k+j}^2 + \sum_{\substack{s=1 \\ s \neq j}}^{l} a_{k+s,k+s} f_s^2 \qquad (2 \le j \le l), \end{aligned}$$

where $\beta \cdot \alpha^2 \equiv \alpha^1 \pmod{m_1}$, and $a_{ij} \equiv a_{k+1,k+1} \equiv a_{k+j,k+j} \equiv 0 \pmod{m_1}$.

Let $A(L)$ be the matrix of the isomorphism L for the chosen bases:

$$A(L) = \begin{pmatrix} 1 & \dots & 0 & a_{1,k+1} & a_{1,k+2} & \dots & a_{1,k+l} \\ & \ddots & & & & & \\ \vdots & 1 & \vdots & \vdots & \vdots & & \vdots \\ & & \ddots & & & & \\ 0 & & 1 & a_{k,k+1} & a_{k,k+2} & \dots & a_{k,k+l} \\ 0 & & 0 & \beta + a_{k+1,k+1} & a_{k+1,k+2} & \dots & a_{k+1,k+l} \\ \vdots & & \vdots & a_{k+2,k+1} & 1 + a_{k+2,k+2} & & a_{k+2,k+l} \\ \vdots & & \vdots & \vdots & & & \vdots \\ 0 & \dots & 0 & a_{k+l,k+1} & \dots & & 1 + a_{k+l,k+l} \end{pmatrix}$$

Without loss of generality we may assume that $\det[A(L)] = 1$; in the opposite case, in place of the basis element f_1^1 we take $-f_1^1$. Since $a_{i,k+j} \equiv 0 \pmod{m_j}$ $(1 \le i \le k+l,\ 1 \le j \le l)$, we have $a_{i,k+j} \equiv 0 \pmod{m_1}$ and $m_j \equiv 0 \pmod{m_1}$.

It is well known that the natural epimorphism $p\colon \mathbf{Z} \to \mathbf{Z}/n\mathbf{Z}$ can be extended to a homomorphism of groups

$$\overline{p}\colon \mathrm{GL}(m, \mathbf{Z}) \to \mathrm{GL}(m, \mathbf{Z}/n\mathbf{Z}).$$

If $A \in \mathrm{GL}(m, \mathbf{Z})$ and $\overline{p}(A)$ is its reduction modulo n, then

$$\det[\overline{p}(A)] = \det A \pmod n. \tag{i}$$

Consider the reduction of $A(L)$ modulo m_1, obtaining

$$\overline{p}(A(L)) = \begin{pmatrix} 1 & & \dots & 0 & 0 & & \dots & 0 \\ & \ddots & & & & & & \\ \vdots & & 1 & \vdots & \vdots & & & \vdots \\ & & & \ddots & & & & \\ 0 & & \dots & 1 & 0 & & \dots & 0 \\ 0 & & \dots & 0 & \rho(\beta) & 0 & \dots & 0 \\ \vdots & & & \vdots & 0 & 1 & & \vdots \\ \vdots & & & \vdots & \vdots & & \ddots & \vdots \\ 0 & & \dots & 0 & 0 & \dots & \dots & 1 \end{pmatrix}$$

On the basis of (i), $p(\beta) = 1 \pmod{m_1}$. Consequently,

$$\beta \cdot \alpha^2 \pmod{m_1} \equiv \beta \pmod{m_1} \cdot \alpha^2 \pmod{m_1} = \alpha^2 \pmod{m_1} = \alpha^1 \pmod{m_1},$$

as was to be proved.

PROOF OF THE LEMMA. Choose in F an arbitrary basis $f_1, \dots, f_k, f_{k+1}, \dots, f_{k+l}$. We have

$$\varphi(f_s) = \sum_{i=1}^{k} a_{is} g_i + \sum_{j=1}^{l} a_{k+j,s} \overline{g}_j, \qquad a_{is} \in \mathbf{Z}$$

$(1 \le s \le k + l)$, where the $a_{k+j,s}$ are residues modulo m_j. Form the square matrix $A(\varphi)$:

$$A(\varphi) = \begin{pmatrix} a_{1,1} & \dots & \dots & a_{1,k+l} \\ \vdots & & & \vdots \\ \vdots & & & \vdots \\ a_{k+l,1} & \dots & \dots & a_{k+l,k+l} \end{pmatrix}$$

We shall reduce $A(\varphi)$ to the desired form by elementary row operations.

Let a_{s1} be the minimal element in the first column. Add (or subtract) a multiple of row s to all nonzero rows, so that in the first column we obtain either 0 or an element less than a_{s1}. Once more, take a minimal element in the first column and carry out an analogous procedure. After a finite number of steps we obtain a basis $\overline{f}_1, \dots, \overline{f}_k, \overline{f}_{k+1}, \dots, \overline{f}_{k+l}$, for which $A(\varphi)$ has the form

$$\overline{A}(\varphi) = \begin{pmatrix} \overline{a}_{11} & \overline{a}_{12} & \dots & \dots & \overline{a}_{1,k+l} \\ 0 & \overline{a}_{22} & \dots & \dots & \overline{a}_{2,k+l} \\ \vdots & & & & \vdots \\ \vdots & & & & \vdots \\ 0 & \overline{a}_{k+l,2} & \dots & \dots & \overline{a}_{k+l,k+l} \end{pmatrix}$$

Since φ is an epimorphism, one can find an element $\overline{\overline{f}}_1$ of the form $\overline{\overline{f}}_1 = \pm\overline{f}_1 + \sum_{j=2}^{k+l} x_j\overline{f}_j$ such that $\varphi(\overline{\overline{f}}_1) = g_1$. Take as a basis $(\overline{\overline{f}}_1, \overline{f}_2, \dots, \overline{f}_k, \overline{f}_{k+1}, \dots, \overline{f}_{k+l})$. The matrix $\overline{A}(\varphi)$ takes the form

$$\overline{\overline{A}}(\varphi) = \begin{pmatrix} 1 & 0 & \dots & \dots & 0 \\ 0 & \overline{a}_{22} & \dots & \dots & \overline{a}_{2,k+l} \\ \vdots & & & & \vdots \\ \vdots & & & & \vdots \\ 0 & \overline{a}_{k+l,2} & \dots & \dots & \overline{a}_{k+l,k+l} \end{pmatrix}$$

By analogous considerations one can find a basis $(\overline{\overline{f}}_1, \dots, \overline{\overline{f}}_k, \overline{f}_{k+1}, \dots, \overline{f}_{k+l})$ for which the matrix of the epimorphism φ is written in the form

$$\tilde{A}(\varphi) = \left(\begin{array}{ccc|ccc} 1 & \dots & 0 & & & \\ \vdots & \ddots & \vdots & & \mathbf{0} & \\ 0 & \dots & 1 & & & \\ \hline & & & \overline{a}_{k+1,k+1} & \dots & \overline{a}_{k+1,k+l} \\ & \mathbf{0} & & \vdots & & \vdots \\ & & & \overline{a}_{k+1,k+1} & \dots & \overline{a}_{k+l,k+l} \end{array}\right)$$

In the following we shall require only the elements $(\overline{f}_{k+1}, \dots, \overline{f}_{k+l})$ of the basis and the lower block. We therefore change notation and put $f_j = \overline{f}_{k+j}$ and $a_{kj} = \overline{a}_{k+j,k+j}$. We obtain the matrix

$$A = \begin{pmatrix} a_{11} & \dots & a_{1l} \\ \vdots & & \vdots \\ a_{l1} & \dots & a_{ll} \end{pmatrix}$$

for the basis $f_1, \dots, f_l$, where the elements of the jth row $(1 \leq j \leq l)$ are residues modulo m_j. Using the previous considerations, one can find a basis $\overline{f}_1, \dots, \overline{f}_l$ for which the matrix A takes the triangular form

$$\overline{A} = \begin{pmatrix} \overline{a}_{11} & 0 & \dots & 0 \\ \overline{a}_{21} & \overline{a}_{22} & & \vdots \\ \vdots & & \ddots & \vdots \\ \overline{a}_{l1} & \dots & \dots & \overline{a}_{ll} \end{pmatrix}$$

Obviously the residues a_{ij} $(1 \leq j \leq l)$ can be considered as elements of the group $\mathbf{Z}_{m_j} = \mathbf{Z}/m_j\mathbf{Z}$. It is clear that $\overline{a}_{ll}$ is a generator of $\mathbf{Z}_{m_l}$: in the opposite case no linear combination of the basis elements $(\overline{f}_1, \dots, \overline{f}_l)$ would be carried to g_l by the mapping φ. If $\overline{a}_{l-1,l-1}$ is a generator of $\mathbf{Z}_{m_{l-1}}$, then we can insert a zero in place of the element $a_{l,l-1}$. We assume that $\overline{a}_{l-1,l-1}$ is not a generator of $\mathbf{Z}_{m_{l-1}}$, and prove the opposite. Since φ is an epimorphism, some linear combination $\gamma = x_1\overline{f}_1 + \dots + x_{l-1}\overline{f}_{l-1} + x_l\overline{f}_l$ of the elements of the basis $\overline{f}_1, \dots, \overline{f}_l$ is carried to g_{l-1} by the mapping φ. It is obvious that $x_l \equiv 0 \pmod{m_l}$; in the opposite case $\varphi(\gamma) \neq g_{l-1}$. We shall prove that $x_l\overline{f}_l = 0$. Consider $x_l\overline{a}_{l,l-1}$; clearly, in the group $\mathbf{Z}_{m_{l-1}}$ the order of $\bar{a}_{l,l-1}$ divides m_{l-1} and consequently divides m_l. Thus we have $m_{l-1} \equiv 0 \pmod{\text{order } \bar{a}_{l-1,l-1}}$, $m_l \equiv 0 \pmod{m_{l-1}}$, and $x_l \equiv 0 \pmod{m_l}$; hence $x_l \equiv 0 \pmod{m_l} \equiv 0 \pmod{m_{l-1}} \equiv 0 \pmod{\text{order } \bar{a}_{l-1,l-1}}$, and consequently $x_l \equiv 0 \pmod{\text{order } \bar{a}_{l-1,l-1}}$; therefore $x_l \bar{a}_{l,l-1} \equiv 0 \pmod{m_{l-1}}$.

One proves analogously that $x_l \cdot \bar{a}_{l_j} \equiv 0 \pmod{m_j}$ for $1 \le j \le l-2$, whence it follows that $x_l \cdot \overline{f}_l = 0$.

Consequently, γ has the form $\gamma = x_1 \overline{f}_1 + \cdots + x_{l-1} \overline{f}_{l-1}$.

It is obvious that $\bar{a}_{l-1,l-1}$ is a generator in $\mathbf{Z}_{m_{l-1}}$, for in the opposite case no linear combination of the elements of the basis $\overline{f}_1, \dots, \overline{f}_{l-1}$ is carried to $\bar{g}_{l-1}$. We can find a basis for which the matrix of the epimorphism φ has the form

$$\overline{\overline{A}} = \begin{pmatrix} \bar{a}_{11} & & & & \\ \bar{a}_{12} & \bar{a}_{22} & & 0 & \\ \vdots & \vdots & \ddots & & \\ \vdots & \vdots & & \bar{a}_{l-1,l-1} & \\ \bar{a}_{l1} & \bar{a}_{l2} & \cdots & 0 & \bar{a}_{ll} \end{pmatrix}$$

By analogous considerations one can show that $\bar{a}_{jj}$ is a generator of $\mathbf{Z}_{m_j}$ $(1 \le j \le l-2)$, and consequently find a basis for which the matrix takes a diagonal form:

$$\tilde{A} = \begin{pmatrix} \bar{a}_{11} & & & & \\ & \ddots & & 0 & \\ & & \ddots & & \\ & 0 & & \ddots & \\ & & & & \bar{a}_{ll} \end{pmatrix}$$

Consider the block

$$\begin{pmatrix} \bar{a}_{l-1,l-1} & 0 \\ 0 & \bar{a}_{ll} \end{pmatrix}.$$

We transform it into the form

$$\begin{pmatrix} \bar{a}_{l-1,l-1} & 1 \\ 0 & \bar{a}_{ll} \end{pmatrix},$$

then to the form

$$\begin{pmatrix} \bar{a}_{l-1,l-1} & 1 \\ \tilde{a}_{l,l-1} & 1 \end{pmatrix},$$

and further we obtain the matrix

$$\begin{pmatrix} \tilde{a}_{l-1,l-1} & 0 \\ \tilde{a}_{l,l-1} & 1 \end{pmatrix}.$$

The considerations which we carried out above show that $\tilde{a}_{l-1,l-1}$ is a generator of $\mathbf{Z}_{m_{l-1}}$. Consequently the matrix takes the form

$$\begin{pmatrix} \tilde{a}_{l-1,l-1} & 0 \\ 0 & 1 \end{pmatrix}.$$

Proceeding upward, we reduce the matrix of the epimorphism φ to the form

$$\begin{pmatrix} \alpha & & & 0 \\ & 1 & & \\ & & \ddots & \\ 0 & & & 1 \end{pmatrix}.$$

If $\alpha > m_1/2$, change α to $-\alpha$ and reduce modulo m_1; we obtain that $\overline{\alpha} = -\alpha \pmod{m_1} < m_1/2$, as was to be proved. The lemma is proved.

We make a remark. Let $\varphi(N)$ be the Euler function, i.e., the number of natural numbers less than N and relatively prime to N. Then there exist $\frac{1}{2}\varphi(m_1)$ inequivalent epimorphisms of free abelian groups of rank $k+l$ onto G.

Let $f\colon M^n \to \mathbf{R}^1$ be an exact Morse function, and put $M^f_\lambda = f^{-1}[0, \lambda + \frac{1}{2}]$. We have a filtration

$$M^f_0 \subset M^f_2 \subset \cdots \subset M^f_\lambda \subset \cdots \subset M^f_n = M^n.$$

The concepts and results cited below can be found in Milnor's book [3].

Let ξ be a gradient-like vector field for f on M^n. It is well known that $H_\lambda(M^f_\lambda, M^f_{\lambda-1}, \mathbf{Z})$ is a free abelian group, generators being represented by left-hand disks corresponding to the vector field ξ. It is also known that for any prescribed basis of the group $H_\lambda(M^f_\lambda, M^f_{\lambda-1}, \mathbf{Z})$ $(2 \le \lambda \le n-2)$ one can find an exact Morse function f' and a gradient-like vector field ξ' for f', coinciding with f and ξ respectively on some neighborhood of $(M^f_{\lambda-1} \cup \overline{M^n - M^f_\lambda})$ and such that the left-hand disks of the vector field ξ', oriented suitably, form the given basis. Fix a minimal system $g^\lambda_1, \dots, g^\lambda_{k_\lambda}, \overline{g}^\lambda_1, \dots, \overline{g}^\lambda_{l_\lambda}$ of generators of the group $H_\lambda(M^n, \mathbf{Z})$, where the g^λ_i are free generators, the $\overline{g}^\lambda_j$ are generators of the torsion, and the order of $\overline{g}^\lambda_j$ is equal to m^λ_j and divides the order of $\overline{g}^\lambda_{j+1}$. From the exact homology sequence of the pair $(M^f_\lambda, M^f_{\lambda-1})$ it follows that $H_\lambda(M^f_\lambda, \mathbf{Z})$ is a free abelian group of rank $k_\lambda + l_\lambda$.

Consider the inclusion of the pair of submanifolds

$$(M^f_\lambda, M^f_{\lambda-1}) \overset{i^f}{\hookrightarrow} (M^n, M^f_{\lambda-1}).$$

We have a commutative diagram

$$\begin{array}{ccccccccc}
0 \to & H_\lambda(M^f_\lambda, \mathbf{Z}) & \overset{j_1}{\to} & H_\lambda(M^f_\lambda, M^f_{\lambda-1}, \mathbf{Z}) & \overset{\partial_1}{\to} & H_{\lambda-1}(M^f_{\lambda-1}, \mathbf{Z}) & \to & H_{\lambda-1}(M^f_\lambda, \mathbf{Z}) & \to 0 \\
& i^\lambda_* \downarrow & \varphi & \downarrow i^f_* & & \downarrow \text{id} & & \downarrow i^\lambda_* & \\
0 \to & H_\lambda(M^n, \mathbf{Z}) & \overset{j_2}{\to} & H_\lambda(M^n, M^f_{\lambda-1}, \mathbf{Z}) & \overset{\partial_2}{\to} & H_{\lambda-1}(M^f_{\lambda-1}, \mathbf{Z}) & \to & H_{\lambda-1}(M^n, \mathbf{Z}) & \to 0
\end{array}$$

$$i^\lambda\colon M^f_\lambda \to M^n$$

The group $j_1(H_\lambda(M^f_\lambda, \mathbf{Z})) \subset H_\lambda(M^f_\lambda, M^f_{\lambda-1}, \mathbf{Z})$ is a direct summand. Construct on M^n an exact Morse function f' and a gradient-like vector field ξ' for f' so that $f = f'$ and $\xi = \xi'$ on $M^n \setminus \operatorname{int}(M^f_\lambda \setminus M^f_{\lambda-1})$, and the left-hand disks of the critical points of index λ of the function f', constructed according to the vector field ξ', give generators of the group $j_1(H_\lambda(M^f_\lambda, \mathbf{Z}))$. It is obvious that

f is equivalent to f'. From the exact homology sequence of the pair (M^n, M_λ^f) it follows that $i_*^\lambda: H_\lambda(M_\lambda^f, \mathbf{Z}) \to H_\lambda(M^n, \mathbf{Z})$ is an epimorphism. Consider the epimorphism $\varphi = i_*^\lambda j_1^{-1}: j_1(H_\lambda(M_\lambda^f, \mathbf{Z})) \to H_\lambda(M^n, \mathbf{Z})$; from the construction it is clear that we arrive at the hypothesis of Lemma 1. Consequently, we can choose f' and ξ' so that the matrix of φ has the form stated in Lemma 1. Thus, to each exact Morse function, for each λ $(2 \le \lambda \le n-3)$ we can uniquely associate a residue α_λ^f.

We call α_λ^f the *λ-residue* of the exact Morse function f.

We have the following theorem.

THEOREM 2. *Exact Morse functions f and g on the manifold M^n are equivalent if and only if all their λ-residues coincide.*

For the proof of Theorem 2 we require the following lemma.

LEMMA 2. *Let f and g be exact Morse functions on M^n so that*

$$f^{-1}[0, \lambda - \tfrac{1}{2}] = g^{-1}[0, \lambda - \tfrac{1}{2}] = M_{\lambda-1}$$

and $\alpha_\lambda^f = \alpha_\lambda^g$. Then there exists a diffeomorphism $h: M^n \to M^n$, isotopic to the identity, such that $h|_{M_{\lambda-1}} = \mathrm{id}$ and $h(f^{-1}[0, \lambda + \frac{1}{2}]) = g^{-1}[0, \lambda + \frac{1}{2}]$ $(0 < \lambda \le n-3)$.

PROOF. Starting from general considerations, one can find a diffeomorphism $\overline{h}: M^n \to M^n$, isotopic to the identity, such that $\overline{h}|_{M_{\lambda-1}} = \mathrm{id}$ and

$$\overline{h}(f^{-1}[0, \lambda + \tfrac{1}{2}]) \subset \operatorname{int} g^{-1}[0, \lambda + \tfrac{1}{2}].$$

In fact, there exist $\varepsilon > 0$ and $\delta > 0$ such that

$$g^{-1}[0, \lambda - \tfrac{1}{2} + \varepsilon] \supset f^{-1}[0, \lambda - \tfrac{1}{2} + \delta].$$

Since $f^{-1}[\lambda - \frac{1}{2}, \lambda - \frac{1}{2} + \delta]$ contains no critical points, $f^{-1}[0, \lambda + \frac{1}{2}]$ has the form

$$f^{-1}[0, \lambda - \tfrac{1}{2} + \delta] \cup_{\psi_1} (D^\lambda \times D^{n-\lambda}) \cdots \cup_{\psi_{k_\lambda + l_\lambda + l_{\lambda-1}}} (D^\lambda \times D^{n-\lambda}),$$

where the $\psi_i: \partial D^\lambda \times D^{n-\lambda} \to f^{-1}[\lambda - \frac{1}{2} + \delta]$ are embeddings. Put $M_{\overline{\lambda}}^g = g^{-1}[0, \overline{\lambda} + \frac{1}{2}]$. Obviously $M_{\overline{\lambda}}^g$ can be represented in the form

$$M_{\overline{\lambda}-1}^g \cup_{\overline{\psi}_1} (D^{\overline{\lambda}} \times D^{n-\overline{\lambda}}) \cdots \cup_{\overline{\psi}_{k_{\overline{\lambda}} + l_{\overline{\lambda}} + l_{\overline{\lambda}-1}}} (D^{\overline{\lambda}} \times D^{n-\overline{\lambda}}),$$

where the $\overline{\psi}_i: \partial D_i^{\overline{\lambda}} \times D^{n-\overline{\lambda}} \to g^{-1}[\lambda + \frac{1}{2}]$ are embeddings $(\lambda < \overline{\lambda} \le n)$. It is obvious that there exists $\tilde{\lambda}$ such that $f^{-1}[0, \lambda + \frac{1}{2}] \subset \operatorname{int} M_{\tilde{\lambda}}^g$. If $(D_i^\lambda \times 0) \cap (0 \times D_j^{n-\tilde{\lambda}}) = \varnothing$, then it is easy to find a diffeomorphism $h_1: M^n \to M^n$, isotopic to the identity, with $h_1|_{M_{\lambda-1}} = \mathrm{id}$, such that

$$h_1(f^{-1}[0, \lambda + \tfrac{1}{2}]) \subset \operatorname{int} M_{\tilde{\lambda}-1}^g.$$

If $(D_i^\lambda \times 0) \cap (0 \times D_j^{n-\tilde{\lambda}}) \ne \varnothing$, then, since $\lambda + n - \tilde{\lambda} < n$, D_i^λ and $D_j^{n-\tilde{\lambda}}$ can be pulled apart, and thus the problem is reduced to the original case. By successively approximating $f^{-1}[0, \lambda + \frac{1}{2}]$ with $M_{\tilde{\lambda}}^g$, we find the desired diffeomorphism $\overline{h}$.

Consider the mappings of pairs, where $M_\lambda^f = f^{-1}[0, \lambda + \frac{1}{2}]$ and $M_\lambda^g = g^{-1}[0, \lambda + \frac{1}{2}]$:

$$\begin{array}{ccc} (M_\lambda^f, M_{\lambda-1}) & \xrightarrow{\;\bar h\;} & (M_\lambda^g, M_{\lambda-1}) \\ & {}_{i^f}\searrow \quad \swarrow_{i^g} & \\ & (M^n, M_{\lambda-1}) & \end{array}$$

We have a commutative diagram

$$\begin{array}{l} 0 \to H_\lambda(M_\lambda^f, \mathbf{Z}) \xrightarrow{j_1} H_\lambda(M_\lambda^f, M_{\lambda-1}, \mathbf{Z}) \xrightarrow{\partial_1} H_{\lambda-1}(M_{\lambda-1}, \mathbf{Z}) \to H_{\lambda-1}(M_\lambda^f, \mathbf{Z}) \to 0 \\ \qquad \bar i_*^f \searrow \qquad\qquad \swarrow \varphi^f \qquad \searrow i_*^f \qquad\qquad\qquad \searrow \\ \bar h_* \Big\downarrow \; 0 \to H_\lambda(M^n, \mathbf{Z}) \to H_\lambda(M^n, M_{\lambda-1}, \mathbf{Z}) \to H_{\lambda-1}(M_{\lambda-1}, \mathbf{Z}) \to H_{\lambda-1}(M^n, \mathbf{Z}) \to 0 \\ \qquad \nearrow \qquad \bar{\bar h}_* \qquad \nwarrow \varphi^g \qquad \nearrow i_*^g \qquad\qquad\qquad \nearrow \\ 0 \to H_\lambda(M_\lambda^g, \mathbf{Z}) \xrightarrow{j_2} H_\lambda(M_\lambda^g, M_{\lambda-1}, \mathbf{Z}) \xrightarrow{\partial_2} H_{\lambda-1}(M_{\lambda-1}, \mathbf{Z}) \to H_{\lambda-1}(M_\lambda^g, \mathbf{Z}) \to 0 \end{array}$$

where $\bar i_*^f\colon M_\lambda^f \to M^n$ and $\bar i_*^g\colon M_\lambda^g \to M^n$ are inclusions. Since $\bar i_*^f$ and $\bar i_*^g$ are epimorphisms, $\bar h_*$ can be either an isomorphism or a monomorphism. If $\bar h_*$ is an isomorphism, then on the basis of Whitehead's theorem (see [4]), $\bar h$ is a homotopy equivalence, since M_λ^f and M_λ^g are simply connected manifolds. Put $Y = (\overline{M_\lambda^g \setminus \bar h(M_\lambda^f)})$. It is easy to prove that $\pi_1(Y) = \pi_1(\partial Y) = 0$. From the exact homology sequence of the pair $[M_\lambda^g, \bar h(M_\lambda^f)]$ it follows that $H_i(M_\lambda^g, \bar h(M_\lambda^f), \mathbf{Z}) = 0$, and so Y is an h-cobordism. According to the h-cobordism theorem, $Y = \partial M_\lambda^g \times I$. Using the I-coordinate, it is easy to find a diffeomorphism $h_1\colon M^n \to M^n$, isotopic to the identity, with $h_1|_{M_{\lambda-1}} = \mathrm{id}$, such that $h_1(\bar h(M_\lambda^f)) = M_\lambda^g$. As h we take the diffeomorphism $h = h_1 \circ \bar h$.

We assume that $\bar h_*$ is a monomorphism, and shall prove that it is an isomorphism. According to the hypotheses of the lemma, f and g have the same λ-residues. Choose basis elements $\{a_i\}$ in $H_\lambda(M_\lambda^f, M_{\lambda-1}, \mathbf{Z})$ and $\{b_i\}$ in $H_\lambda(M_\lambda^g, M_{\lambda-1}, \mathbf{Z})$ $(1 \le i \le k_\lambda + l_\lambda)$ such that the matrices of the epimorphisms $\varphi^f = i_*^f \circ j_1^{-1}$ and $\varphi^g = i_*^g \circ j_2^{-1}$ have the form of Lemma 1. Then it is obvious that $\bar h_*(a_i) = b_i + \sum_{j \ne i} \theta_{ij} b_j$ $(1 \le i \le k_\lambda + l_\lambda)$, where $\theta_{ij} \equiv 0 \pmod{m_j}$. Without loss of generality, we may suppose that g is chosen so that

$$(\partial D_j^{\lambda+1} \times 0) \cap (0 \times D_i^{n-\lambda}) = m_j \delta_{ij},$$

where the $D_j^{\lambda+1} \times 0$ are the left-hand disks of the critical points of index $\lambda + 1$ (see [5]). Using a general position argument and the Whitney lemma (see [3]), one can obtain $(D_i^\lambda \times 0) \cap (0 \times \overline{D}_j^{n-\lambda}) = \delta_{ij}\bar\theta_{ij}$, where $\bar\theta_{ij} = \theta_{ij}$ for $1 + k_\lambda \le j \le l_\lambda$ and $\bar\theta_{ij} = 0$ for $1 \le j \le k_\lambda$ $(i, j = 1, \ldots, k_\lambda + l_\lambda)$, where $(D_i^\lambda \times 0)$ is the left-hand disk of the critical point of the function f corresponding to the generator a_i, and $(0 \times \overline{D}_j^{n-\lambda})$ is the right-hand disk of the critical point of the function g corresponding to the generator b_j. Using the connected sum of $D_i^\lambda \times 0$ with $\partial(D_j^{\lambda+1} \times 0)$, one can decrease (or increase) the number of points of intersection of $(\partial D_i^\lambda \times 0)$ with $(0 \times \overline{D}_j^{n-\lambda})$ by m_j $(1 \le j \le l_\lambda)$. In fact, let $\tilde h\colon M^n \to M^n$ be a diffeomorphism, isotopic to the identity, with

$$\tilde{h}(\partial D_j^{\lambda+1} \times 0) \subset \operatorname{int} M_\lambda^g, \qquad \tilde{h}(\partial D_i^\lambda \times 0) \cap (D_i^\lambda \times D_i^{n-\lambda}).$$

Consider the connected sum

$$A_i = (D_i^\lambda \times 0) \# \tilde{h}(\partial D_j^{\lambda+1} \times 0).$$

Since A_i is a disk, by taking its normal bundle we obtain a new handle, glued according to the old mapping, and consequently $A_i \cap (0 \times D_j^{n-\lambda}) = \theta_{ij} \pm m_j$. Repeating this construction a finite number of times, we obtain that $\theta_{ij} = 0$. Consequently $\overline{h}_*$ is an isomorphism. The lemma is proved.

PROOF OF THEOREM 2. The proof repeats the considerations of Agoston. On the basis of [6], one can find a diffeomorphism $h_{i_1}: M^n \to M^n$, isotopic to the identity, with $h_{i_1}(M_0^f) = M_0^g$. Assume by induction that we have found diffeomorphisms $\{h_{i_s} | s = 1, \dots, k\}$ satisfying the following conditions:

1) $h_{i_s} \approx h_{i_{s-1}}$.
2) $h_{i_s}|_{M^1_{i_s-1}} = h_{i_{s-1}}|_{M^1_{i_s-1}}$.
3) $h_{i_s}(M^1_{i_s}) = M^2_{i_s}$.

Let $\eta_s = fh_{i_k}^{-1}$ and $M_i^{\eta_s} \equiv \eta_s^{-1}[0, i + \frac{1}{2}]$. Then $M_{i_k}^{\eta_s} = M_{i_k}^g$. Assume that $2 \le i_k \le n-2$. Applying Lemma 2 to η_s and g, we find a diffeomorphism $h: M^n \to M^n$, isotopic to the identity and with $h(M^{\eta_s}_{i_{k+1}}) = M^g_{i_{k+1}}$; then put $h_{i_{k+1}} \equiv h \circ h_{i_k}$. It is obvious that $h_{i_{k+1}}$ satisfies conditions 1)–3). If $n-2 \le i_{k+1} \le n$, then by applying analogous considerations to the functions g and $n - \eta_s$, we obtain the desired diffeomorphism. The theorem is proved.

Analogous theorems are valid in the PL and Top categories.

REFERENCES

1. S. Smale, *On the structure of manifolds*, Amer. J. Math. **84** (1962), 387–399.

2. Max K. Agoston, *On handle decompositions and diffeomorphisms*, Trans. Amer. Math. Soc. **137** (1969), 21–26.

3. John Milnor, *Lectures on the h-cobordism theorem*, Princeton Univ. Press, Princeton, N.J., 1961.

4. Edwin H. Spanier, *Algebraic topology*, McGraw-Hill, 1966.

5. V. V. Sharko, *Smale's theorem for simply connected topological manifolds*, Mathematics Collection (Yu. A. Mitropol′skiĭ, editor), "Naukova Dumka", Kiev, 1976, pp. 119–121. (Russian)

6. John Milnor, *Differentiable structures*, mimeographed notes, Princeton Univ., Princeton, N.J., 1961.

Translated by P. S. LANDWEBER

Amer. Math. Soc. Transl.
(2) Vol. **134**, 1987

On an Algorithm for Successive Improvement of a Strategy

UDC 681.3:519.21

V. V. BARANOV AND N. S. PODTSYKIN

A system whose evolution can be described by a random process is said to be stochastic. A stochastic system is said to be a controllable Markov process if it can be described by a set of objects $\{E, Y, Q, \Pi, R\}$, where E is a set of states, Y is a set of controls, $Q = Q(E|E \times Y)$ is a set of conditional probability measures determining the transitions on E under the action of controls $y \in Y$, Π is a set of rules for selecting the controls, and $R = R(E \times Y \times E)$ is a set of consequences of the choice of controls, expressed as penalties, income, or payoffs.

The problem is to determine a rule $\pi^* \in \Pi$ for selecting controls (called strategies in what follows) having the most favorable consequences. We confine ourselves to the class of nonrandomized stationary Markov strategies. Let

$$g(x, \pi) = \lim \frac{1}{n+1} M_x^\pi \sum_{k=0}^{n} w(x_k, y_k, x_{k+1}) \tag{1}$$

be the average expected payoff for a strategy π and an initial state $x \in E$, where $w(x_k, y_k, x_{k+1})$ is the immediate payoff in passing from $x_k \in E$ to $x_{k+1} \in E$ under the action of a control $y_k \in Y$.

In principle, an optimal strategy can be found with the help of an algorithm of Howard [1] or by reduction to a linear programming problem [2]. However, in each of these methods it becomes necessary to solve systems of linear equations, which makes the problem considerably more difficult, and even impossible to solve in practice when the dimensions are large.

In this article we present an algorithm for finding an optimal strategy by successively improving the initial strategy. It is not necessary to solve systems of linear equations in this algorithm, and that enables us to solve problems of practically any dimension.

1980 *Mathematics Subject Classification* (1985 *Revision*). Primary 60J10, 65V05, 93E20.

Translation of Vestnik Khar'kov. Univ. No. 134 (Mat. Mekh. Vyp. (4) 41) (1977), 16–23; MR **58** #9731.

The set of states and the set Y of controls are assumed to be finite below. We need some facts from the theory of finite Markov chains [3].

LEMMA. *Let $T = T(\pi)$ be an N-dimensional regular transition matrix. Then:*
a) *T is Cesàro summable to a matrix T^*, i.e.,*

$$\lim_{n\to\infty} \frac{1}{n+1} \sum_{k=0}^{n} T^k = T^*.$$

Moreover, the powers T^k converge to T^* as $k \to \infty$.
b) *T^* has the property that*

$$TT^* = T^*T = T^*T^* = T^* \tag{2}$$

and is the matrix of final probabilities of the states, with its columns all equal to the same vector $t = \{t_x,\ x \in E\}$, i.e., $T^* = et$, where e is the N-dimensional column vector with all elements equal to 1. The vector t does not depend on the initial probability distribution of the states and is the unique solution of the system

$$t = tT, \qquad t_x \geq 0, \quad x \in E, \qquad te = 1.$$

c) *The matrix $A = I - (T - T^*)$ is nonsingular, i.e., the matrix $Z = A^{-1} = \sum_{k=0}^{\infty}(T - T^*)^k$ exists.*
d) *Let $H = \sum_{k=0}^{\infty}(T^k - T^*)$. Then $H = Z - T^*$ and*

$$(I - T)H = H(I - T) = I - T^*. \tag{3}$$

Suppose now that the column vector $w = w(x, y, x)$ of immediate payoffs and the transition matrix $T = T(\pi)$ are given. Let $T_x(\pi)$ denote the xth column of $T(\pi)$, and $w(\pi)$ the payoff vector with components

$$w(x, \pi(x)) = T_x(\pi)w = \sum \alpha_{xz}(\pi(x))w(x, \pi(x), z),$$

where α_{xz} are the elements of the row $T_x(\pi)$.

Let $G(\pi)$ be the vector with components $g(x, \pi)$. Obviously,

$$G(\pi) = \lim_{n\to\infty} \frac{1}{n+1} \sum_{k=0}^{n} T^k(\pi)w(\pi).$$

The components of $G(\pi)$ have the form $g(x, \pi) = t(\pi)w(\pi)$, where $t(\pi)$ is the vector of final probabilities of the states. This vector does not depend on the initial state. Therefore, $g(x, \pi) = g(\pi)$ for all $x \in E$. Then $G(\pi)$ can be written as $G(\pi) = g(\pi)e$, where e is the column vector with components equal to 1.

THEOREM 1. *Let $w = w(\pi)$ be the payoff vector, $T = T(\pi)$ the transition matrix, and $T^* = T^*(\pi)$ the final matrix. Then the average payoff $g = g(\pi)$ satisfies the following relation, written in vector form:*

$$ge + r = w + Tr, \tag{4}$$

where

$$ge = T^*w, \qquad r = Hw. \tag{5}$$

PROOF. By (3),

$$(I - T)Hw = (I - T^*)w,$$

whence

$$T^*w + Hw = w + THw.$$

Using the notation (5), we get the required relation.

REMARK. The vector r is relative, i.e., the vector $r+ce$, where c is an arbitrary constant, also satisfies (4). This is not hard to check by substitution.

THEOREM 2. *Let f and π be given strategies, and let $T(f)$ and $T(\pi)$ be the regular matrices corresponding to them. If the vector inequality*

$$(T(f) - T(\pi))r(\pi) \geq w(\pi) - w(f) \tag{6}$$

holds, then

$$g(f) \geq g(\pi). \tag{7}$$

PROOF. Using (6), we get an N-dimensional column vector $\gamma(f)$:

$$\gamma(f) = w(f) + T(f)r(\pi) - w(\pi) - T(\pi)r(\pi) \geq 0. \tag{8}$$

Let us write equation (4) for the strategies f and π:

$$g(f)e + r(f) = w(f) + T(f)r(f),$$
$$g(\pi)e + r(\pi) = w(\pi) + T(\pi)r(\pi).$$

Subtracting the last equality from the first and considering (7) and (8), we get

$$\begin{aligned} g(f)e - g(\pi)e + r(f) - r(\pi) &= w(f) + T(f)r(f) - w(\pi) - T(\pi)r(\pi) \\ &= \gamma + T(f)[r(f) - r(\pi)]. \end{aligned}$$

Let $\Delta g = g(f) - g(\pi)$ and $\Delta r = r(f) - r(\pi)$; then the last equality can be written in the form

$$\Delta ge + \Delta r = \gamma + T(f)\Delta r.$$

We multiply both sides of this equality from the left by the final vector $t(f)$,

$$t(f)\Delta ge + t(f)\Delta r = t(f)\gamma + t(f)T(f)\Delta r;$$

but $t(f)T(f) = t(f)$, so $t(f)\Delta ge + t(f)\Delta r = t(f)\gamma + t(f)\Delta r$. Therefore, $t(f)\Delta ge = t(f)\gamma$. Since the components of the vector γ are nonnegative, while those of $t(f)$ are positive, $t(f)\gamma \geq 0$. Hence, $t(f)\Delta ge = \Delta g \geq 0$, which proves the theorem.

REMARK. It is not hard to see that if there is at least one component for which strict inequality holds in (6), then $g(f) > g(\pi)$.

It follows from Theorem 2 that to find a strategy f satisfying $g(f) > g(\pi)$ it is necessary to know the vector

$$r(\pi) = Hw = \sum_{k=0}^{\infty}(T^k w - T^* w).$$

We show that $r(\pi)$ can be represented in the form

$$r(\pi) = \sum_{k=0}^{\infty}(T^k w - T_z^k w),$$

where $T_z^k w$ is the zth component of $T^k w$. Indeed, suppose that the constant c is such that $T^*(w + ce) = 0$. Since T^* and T^k $(k = 0, 1, \ldots)$ are stochastic matrices,

$$\begin{aligned} r(\pi) = \sum_{k=0}^{\infty}(T^k w - T^* w) &= \sum_{k=0}^{\infty}(T^k(w + ce) - T^*(w + ce)) \\ &= \sum_{k=0}^{\infty} T^k(w + ce). \end{aligned}$$

It is known that for the regular matrix T each component of r is representable by an absolutely convergent series [3]. Therefore,

$$\sum_{k=0}^{\infty} T_z^k(w + ce) = a < \infty.$$

Consider the vector

$$\begin{aligned} r - ae &= \sum_{k=0}^{\infty} T^k(w + ce) - \sum_{k=0}^{\infty} T_z^k(w + ce) \\ &= \sum_{k=0}^{\infty}(T^k w + T^k ce - T_z^k w - T_z^k ce) = \sum_{k=0}^{\infty}(T_w^k - T_2^k w). \end{aligned}$$

The vector $r - ae$ satisfies, along with r, the relation (4). Therefore, we set $r = \sum_{k=0}^{\infty}(T^k w - T_z^k w)$, which is what was required to prove.

Let

$$r_n(\pi) = \sum_{k=0}^{n}(T^k(\pi)w - T_z^k(\pi)w). \tag{9}$$

Hence, $r(\pi) = \lim_{n\to\infty} r_n(\pi)$. Then the xth component of inequality (8) can be written as

$$\lim_{n\to\infty} \gamma_n(x, y) = \gamma(x, y) \geq 0, \tag{10}$$

where

$$\gamma_n(x, y) = (T_x(y) - T_x(\pi))r_n(\pi) - w(x, \pi) + w(x, y), \quad y = f(x). \tag{11}$$

Theorem 2 serves as a basis for constructing an algorithm for synthesizing an optimal strategy.

Let $Y_x \subseteq Y$ be the set of admissible controls in the state $x \in E$. It follows from (8) that to find a control $y \in Y_x$ improving a strategy in the state $x \in E$ it is necessary to establish the sign of the quantity $\gamma(x, y)$. We use the representation (10) and estimate the sign of γ from a finite number of terms in the sequence $\{\gamma_k\}$. The number of terms of the sequence ensuring a sufficiently good approximation γ_k of the limit γ can be estimated from the size of the difference $\gamma_k - \gamma_{k-1}$

between neighboring terms. Obviously, if this difference stays sufficiently small in modulus beginning with some index m, then the terms γ_k, $k > m$, of the sequence are close to their limit. Although $\{\gamma_k\}$ is a convergent sequence, it does not converge monotonically in general. Therefore, to estimate the closeness of the terms γ_k to their limit, with the purpose of diminishing the possibility of erroneous solutions due to the fact that $\{\gamma_k\}$ is not a monotone sequence, it is advisable to use the differences $\overline{\gamma}_k - \overline{\gamma}_{k-1}$ between the corresponding arithmetic means instead of $\gamma_k - \gamma_{k-1}$, where

$$\overline{\gamma}_k = \frac{1}{k+1}\sum_{i=0}^{k}\gamma_i.$$

Then to choose the number k^* of terms in the sequence for which sufficient closeness to the limit is attained it is natural to use the condition

$$k^* = \min\{k\colon |\overline{\gamma}_k - \overline{\gamma}_{k-1}| \le \varepsilon\}, \tag{12}$$

where $\varepsilon > 0$ is a sufficiently small fixed constant.

To choose a control improving a strategy it now suffices to use the sign of the term γ_{k^*}: if $\gamma_{k^*}(y) > 0$, then y is taken as a control improving the strategy, and if $\gamma_{k^*}(y) \le 0$, then the given control y is rejected.

The rule described for choosing a control improving a strategy is realized for a given fixed strategy π and the corresponding vector $r(\pi)$.

We remark that the vector $r_k(\pi)$ is computed, and $r(\pi)$ is thereby estimated, at the same time that the vector γ is estimated.

To avoid new computations of the vectors $r_k(\pi)$, $k = 1, \ldots, k^*$, in testing the next control in a state $x \in E$ it is obviously advisable to use the computed value of $r_{k^*}(\pi)$ and choose all controls satisfying (6), i.e., improving the strategy, according to the rule described.

Let

$$\gamma^*_{k^*}(x) = \max_{y \in Y_x} \gamma_{k^*}(x, y). \tag{13}$$

Then one iteration in the procedure consists in performing the following operations: compute $\gamma^*_{k^*}(x)$ for all $x \in E$ for a given strategy π and a given $\varepsilon > 0$, choose those controls for which $\gamma^*_{k^*}(x) > 0$ as controls improving the strategy, and from the chosen controls at each $x \in E$ compose a new strategy π' which is an improved strategy. This finishes one iteration, and the procedure is repeated.

To prevent cycling in condition (12) it is expedient to decrease ε at each iteration, say by setting it equal to $\varepsilon(h) = \varepsilon^h$, where h is the number of the iteration.

It is not hard to see that our algorithm converges to an optimal strategy in finitely many iterations. This follows from the fact that the set of strategies is finite, and the error in the choice of an improved strategy decreases at each iteration.

The search for an optimal strategy is stopped as soon as the same strategy is obtained in two successive iterations.

Thus, the algorithm for successive improvement of a strategy consists in the following.

1. Choose a strategy π as an initial approximation.

2. For each $x \in E$ compute the quantities $\gamma_k(x, y)$, $y \in Y_x$, $k = 0, 1, \dots, k^*$, where

$$k^* = \min\{k\colon |\overline{\gamma}_k - \overline{\gamma}_{k-1}| \le \varepsilon^h\}$$

with $0 < \varepsilon < 1$, $\overline{\gamma}_k = (1/(k+1)) \sum_0^k \gamma_i$, and h the number of the iteration.

3. For each $x \in E$ compute

$$\gamma^*_{k^*}(x) = \max_{y \in Y_x} \gamma_{k^*}(x, y).$$

4. Compose a strategy from the controls y such that $\gamma^*_{k^*}(x) > 0$.

5. If the strategy $f(h) = f(h-1)$, then stop; if $f(h) \ne f(h-1)$, then go back to step 1.

EXAMPLE. Suppose that we have a system which can be found in one of three states. The sets Y_x, $x = 1, 2, 3$, contain the following controls: $Y_1 = \{y_{11}\}$, $Y_2 = \{y_{21}, y_{22}, y_{23}\}$, $Y_3 = \{y_{31}\}$.

Application of a control $y \in Y_x$ in the state x leads to a transition of the system to a state j with probability $\alpha_{xj}(y)$. Here we obtain an income $w(x, y)$. Suppose that the following transition probability and income vectors are given:

$$\begin{aligned}
\alpha_{1j}(y_{11}) &= \{0.2\ 0.5\ 0.3\}, & w(1, y_{11}) &= 0,\\
\alpha_{2j}(y_{21}) &= \{0.4\ 0.2\ 0.4\}, & w(2, y_{21}) &= 1,\\
\alpha_{2j}(y_{22}) &= \{0.2\ 0.3\ 0.5\}, & w(2, y_{22}) &= 0.8,\\
\alpha_{2j}(y_{23}) &= \{0.5\ 0.2\ 0.3\}, & w(2, y_{23}) &= 1.3,\\
\alpha_{3j}(y_{31}) &= \{0.4\ 0.4\ 0.2\}, & w(3, y_{31}) &= 2.
\end{aligned}$$

1. An initial strategy π is specified by choosing y_{11}, y_{21}, and y_{31}. Then

$$T(\pi) = \begin{pmatrix} 0.2 & 0.5 & 0.3 \\ 0.4 & 0.2 & 0.4 \\ 0.4 & 0.4 & 0.2 \end{pmatrix}, \qquad w(\pi) = \begin{pmatrix} 0 \\ 1 \\ 2 \end{pmatrix}.$$

2. Let $z = 1$ and $\varepsilon = 0.1$. Compute the vector $r_0(\pi)$ and $r_1(\pi)$. Then, by (11), for each control $y \in Y_2$ (except y_{21}, since it is in π) computation of γ_0 and γ_1 yields

$$\gamma_0(2, y_{22}) = 0.1; \quad \gamma_1(2, y_{22}) = 0.06; \quad \gamma_0(2, y_{23}) = 0.1; \quad \gamma_1(2, y_{23}) = 0.13.$$

Check condition (12) for y_{22},

$$|\tfrac{1}{2}0.16 - 0.1| = 0.02 < \varepsilon - 0.1,$$

and for y_{23},

$$|\tfrac{1}{2}0.23 - 0.1| = 0.015 < \varepsilon.$$

From this we get $k^* = 1$.

3. $\gamma^*_{1^*}(2) = 0.13 > 0$.

4. Compose a strategy f from the controls y_{11}, y_{23}, and y_{31}.

5. Proceed to the second iteration, with f as the initial strategy.

2. In a way similar to the preceding, we obtain

$$\gamma_0(2, y_{21}) = -0.1; \quad \gamma_1(2, y_{21}) = -0.143.$$

The condition (12) fails to hold for $k = 1$. Compute $\gamma_2(2, y_{21}) = -0.1258$. The condition (12) holds for $k = 2$; namely,

$$|-\tfrac{1}{3}0.3688 + \tfrac{1}{2}0.243| = 0.0014 < \varepsilon^2 = 0.01$$

Next, we find

$$\gamma_0(2, y_{22}) = 0.03; \quad \gamma_1(2, y_{22}) = -0.075; \quad \gamma_2(2, y_{22}) = -0.0489.$$

Condition (12) holds for $k = 2$; namely,

$$|-\tfrac{1}{3}0.0939 + \tfrac{1}{2}0.045| = 0.0088 < \varepsilon^2$$

and $k^* = 2$.

3. $\gamma^*_{2^*}(2) = -0.0489 < 0$.

4. Since $\gamma^*_{2^*} < 0$, the strategy was not improved at the second iteration.

5. The strategy $f(2)$ coincides with $f(1)$ and is optimal.

Our algorithm is of important practical significance, since it enables us to efficiently solve many problems in operations research, for example, in the preventative inspection of complicated systems, inventory control, etc.

BIBLIOGRAPHY

1. Ronald A. Howard, *Dynamic programming and Markov processes*, M.I.T. Press, Cambridge, Mass., and Wiley, New York, 1960.

2. Alan S. Manne, *Linear programming and sequential decisions*, Management Sci. **6** (1960), 259–267.

3. John G. Kemeny and J. Laurie Snell, *Finite Markov chains*, Van Nostrand, Princeton, N.J., 1960.

Translated by H. H. McFADEN

ABCDEFGHIJ — 8987